Technoscience and Postphenomenology

Postphenomenology and the Philosophy of Technology

Technological advances affect everything from our understandings of ethics, politics, and communication, to gender, science, and selfhood. Philosophical reflection on technology helps draw out and analyze the nature of these changes, and helps us understand both the broad patterns and the concrete details of technological effects. This book series provides a publication outlet for the field of the philosophy of technology in general, and the school of thought called "postphenomenology" in particular. Philosophy of technology applies insights from the history of philosophy to current issues in technology, and reflects on how technological developments change our understanding of philosophical issues. In response, postphenomenology analyzes human relationships with technologies, while integrating philosophical commitments of the American pragmatist tradition of thought.

Postphenomenology and Technoscience: The Manhattan Papers, edited by Jan Kyrre Berg O. Friis and Robert P. Crease

Postphenomenological Investigations: Essays on Human–Technology Relations, edited by Peter-Paul Verbeek and Robert Rosenberger

Design, Meditation, and the Posthuman, edited by Dennis M. Weiss, Amy D. Propen, and Colbey Emmerson Reid

Technoscience and Postphenomenology

The Manhattan Papers

Edited by
Jan Kyrre Berg O. Friis
and Robert P. Crease

LEXINGTON BOOKS
Lanham • Boulder • New York • London

Published by Lexington Books
An imprint of The Rowman & Littlefield Publishing Group, Inc.
4501 Forbes Boulevard, Suite 200, Lanham, Maryland 20706
www.rowman.com

Unit A, Whitacre Mews, 26-34 Stannary Street, London SE11 4AB

British Library Cataloguing in Publication Information Available

Library of Congress Control Number: 2015949370
ISBN: 978-0-7391-8961-0 (cloth : alk. paper)
eISBN: 978-0-7391-8962-7

∞™ The paper used in this publication meets the minimum requirements of American National Standard for Information Sciences—Permanence of Paper for Printed Library Materials, ANSI/NISO Z39.48-1992.

Printed in the United States of America

Contents

Foreword: Is There a Bat problem for Postphenomenology? vii
Don Ihde

Introduction xvii

1 Postphenomenology's North American Future 1
Robert C. Scharff

2 Beyond Originary Givenness? Postphenomenology, Digital Imaging and Evidentiary Responsibility 19
Shannon Vallor

3 Historical Variations and the Cellular Age 39
Galit Wellner

4 On Postphenomenology and the Postcolonial 57
Srikanth Mallavarapu

5 What is Multistability? A Theory of the Keystone Concept of Postphenomenological Research 69
Kyle Powys Whyte

6 Hospital Architecture and Design in Postphenomenological Perspective 83
Lars Botin

7 Post-Telescope-Postphenomenology? . . . and a Little Locomotive History 105
Michael Funk

8 The Dubstep Mashup 115
Stacey O. Irwin

9 Postphenomenology: What's New? What's Next? 129
Robert Rosenberger

10 An Introduction to Hyperology: The Age of the Chimera 149
Roisin Lally

11 Nursing's Nightingale Needed a Lamp! 161
Anette Forss

12 Multistable Roboethics 169
Cathrine Hasse

13 Toward a Theory of Technological Mediation: A Program for Postphenomenological Research 189
Peter-Paul Verbeek

14 Don Ihde's Relevance to the Gun Debates 205
Evan Selinger

15 Technology and the Environment: Lessons from Fukushima 211
Junichi Murata

16 How Does Technology Alter Sports: Body, Space, and Ethics 225
Shoji Nagataki

17 Somatology of Aurality: The Voice of the Material World 239
Eduardo Mendieta

Index 249

About the Contributors 255

Foreword

Is There a Bat problem for Postphenomenology?

Don Ihde

Thomas Nagel's famous article, "What is it like to be a bat?" was published in 1974.[1] In context it was an argument against physicalist reductionism. Nagel was arguing that conscious experience—subjectivity—could not be reduced to brain states, but in the process also concluded that we could not know what it was like to be a bat. That year was crucial for me in a phenomenological sense—I had been working on my book, *Listening and Voice: A Phenomenology of Sound* (1976, first edition, later a second edition, 2007). In the process and during writing from 1972–1974, I was teaching myself to echo-locate. This was to lead to a set of auditory experiments which I used to teach my phenomenology students to also learn a degree of echo-location in the halls of Harriman Hall at Stony Brook.

Those familiar with the history of twentieth-century philosophies will be aware that both phenomenology (Husserl) and pragmatism (Dewey) were busy trying to deconstruct early modern epistemology (Descartes and Locke, seventeenth century). At the turn of the millennium, *Nature* magazine invited me to do a millennium essay, which I did with the title, "Epistemology Engines" (2000).[2] In that essay I showed how both Descartes and Locke separately, but deliberately, modeled the birth of early modern epistemology on the *camera obscura,* a favorite optical device of the Renaissance and early modernity. This was to be the invention of "subjectivity" and "objectivity," which modeled the eye to mind to a homunocular subject in the box, viewing mental events. Knowledge was the correspondence of an event inside and an object outside the camera box, and with it experience becomes "subjective" and private.

As I reread Nagel, I wondered how far he had escaped the seventeenth century. Here is what I found: unlike Descartes, Nagel was generous to conscious subjectivity well beyond humans, "Conscious experience is a widespread phenomenon. It occurs at many levels of animal life, though we cannot be sure of its presence in the simpler organisms . . . ."[3] And he eschews the crudeness of the homunocular subject-in-a-box, "I am not averting here to the alleged privacy of experience to its possessor"[4] although I worry that there is a vestige of that position in his insistence that "the facts of experience . . . are accessible only from one point of view."[5] And, finally, although he is arguing against physicalist reductionism, he gives back what he takes away by averting to brain structures which are different in bats and humans, thus making it impossible for us to fully imagine bat being. "Their brain structure is different."[6]

The other side of the argument refers to the subjective nature of experience which in the article displays considerable ambivalence. Nagel's argument relies on how close bat subjectivity might be to human subjectivity and since echo-location is obviously more refined with bats, we apparently are not able to approximate or imagine into such a subjectivity (privacy again?). Nagel does avert to heightened auditory abilities of skilled blind persons in a footnote[7] but all this remains short of bat subjectivity. So, while I conclude Nagel has become more sophisticated than an early modern, he has done so only to a limited degree. But I now want to leave Nagel and turn to phenomenology and postphenomenology, using the bat problem as an index for a difference between these two versions of phenomenology.

If Dewey and Husserl began the attack upon early modern epistemology early in the twentieth century, Maurice Merleau-Ponty dealt it a *coup-de-grace* by mid-century. "This phenomenal field is not an 'inner world,' the 'phenomenon' is not a 'state of consciousness' or a 'mental fact' and the experience of phenomena is not an introspection."[8] How far is this from the "subjectivity" of a single alien consciousness of a Nagel bat? "To be a consciousness, or rather *to be an experience* is to have an inner communication with the world, the body, and others, to be with them rather than beside them."[9] Here we have a quite different set of descriptions from those of early modernity. I am taking Merleau-Ponty as a classic phenomenologist who takes a motile whole body approach to experienced embodiment. It should be noted that already here phenomenology no longer fits into the primacy of subjectivity and consciousness. Embodiment is actional, more than mere consciousness, and ontologically interrelational, in-the-world. Later I will point to Merleau-Ponty's equally suggestive postphenomenological anticipations through instrumental material means. But for the moment I would like to follow a trajectory which links what I shall call *connoisseur perception* to classical phenomenology.

Connoisseur perception is skilled, trained perception. Here are a few simple variations:

- Wine tasting is such a skill. There has been a more than fifty-year tradition of a wine tasting contests between teams from Oxford and Cambridge Universities in the UK. Chosen teams of student tasters take up rigorous training for several months before the competitive event. They learn to detect the wine variety (most oenophiles can easily tell the difference between a cabernet, merlot, malbec, etc), vintage, region, and multiple distinct flavors related to the tannin, acid, sugar contents of wine materiality. This is trained, connoisseur perception, an embodiment skill.
- More ordinary—bird watching. There is a vast variety of finch and sparrow species, all of which are likely to be perceived simply as small, brown, birds by the amateur. But with trained perception, one can soon tell the different markings on the bird's heads, throats, wing patterns, beak patterns and the like—plus the distinctive songs which differ greatly between song sparrows, white throated sparrows, chipping sparrows and the like, recognizable to connoisseur perception, the embodiment skill for appropriate phenomenological observation.
- Or, one of my own favorites is mushroom hunting. My wife and I recognize a varied set of Vermont mushrooms for warm weather eating: oysters, chanterelles, boletus, puff, etc. etc. We are amused that some guests are afraid to sample, but others find fresh trout and chanterelles a special delight.

Classical phenomenology, I claim, reaches these levels of experiential analysis. All the examples have been of unaided or unmediated perceptual-bodily experience—embodiment. And here I can now return to the bat problem. Auditory connoisseur perception can easily recognize a range of human echo-location. Nagel recognized some skills attained particularly by blind auditory connoisseurs—I have seen some documentaries of blind bicycle riders. These riders produce a mouthed click sound (clicks or sounds of short temporal span are best for echo-location. Longer sounds as in songs are different, see below). These riders are able to detect poles and other objects down to three inch diameters. In *Listening and Voice* I have a chapter on perceiving shapes, surfaces, and interiors, all aspects of human echo-location. One can auditorily determine spatial-material features. A shoe box in which one can place glass marbles, or gummy-bear marbles, acoustically sound different spatial-material features. Pounding on walls can reveal where a stud is. A granite countertop sounds different from a wooden one, etc. All of this is material-spatial as is the more finely tuned bat version.

I also try to keep up with much frontier work in acoustic and auditory discovery, and two bat trivia are worth citing: in Central America small tungara frogs are prey for night feeding bats. These frogs sing mating songs hoping to attract females—but the frogs can hear bat echo-location clicks and so stop their songs. But it is too late, because the sound waves from the songs have produced ripples on the puddles they frequent and the bat can "read" the wave pattern to find the frog![10] Or, a more interesting bat-flower interaction is that of certain flowering plants, which change shape at night to form a shape that can reflect the echoes of bat clicks so they may be discovered and pollinated. This is another exquisite example of interrelational adaptation.[11]

These examples of connoisseur perception are all excellent variants upon Merleau-Ponty's notion of perceptual-bodily motility as part of the mutually human-world interrelationality. I now want to begin the extension to postphenomenology which in two further steps *incorporates* two different types of bodily perception by taking "technologies" or specifically in this case instrumental technologies into *intentionality* or phenomenological interrelationality. I shall call what results *instrumentally enhanced* and *instrumentally translational perception.* Again, Merleau-Ponty may be seen to have anticipated this move. I have previously referred to his example of an extended sense of embodiment or bodily environmental experience with the feathered hat of a lady and of the blind man's cane. In both these cases the experience is mediated through a material artifact, and gives the human experiencer a perception of extended embodiment. Mediating instrumentation, however, gives a more precise example of this now enhanced and extended perception.

I will begin with my hearing prosthesis, state-of-the-art digital hearing aids. These are designed to enhance speech. As already noted, sounds may be short and click-like, in speech—consonants. And sounds may be longer and tone or whistle-like—vowels. Both need to be perceived to comprehend speech—but while it is easy to simply amplify vowels, only digital processes can manipulate and amplify consonants. So, what my acoustic prosthesis does is to enhance and constitute these sound components so I may understand speech. This is an example of instrumentally enhanced perception which remains isomorphically acoustic. Telescopes and microscopes as optical instruments did this for early modernity for vision. Today, of course, we have much higher degrees of enhancement—take for example, night vision goggles which enhance ambient light and make it possible for humans to "see" objects in the dark (albeit sort of greenish in color). Or, one may find many examples of both visual and acoustic imaging which pushes all previous magnificational or amplificational limits. Electronic and scanning tunneling microscopes for micro phenomena, multilensed and compound telescopes for macro phenomena are now deployed. These make it possible for enhanced bodily perceptual experience never previously possible and it is to this human-instrument-world interrelationship that postphenomenology can and does attend.

What I am here calling instrumentally translational perception is yet another mode of pushing horizons. I have frequently pointed out that all astronomy—until definitively in the twentieth century—was optical. Ancient "eyeball" observations were revolutionized by instrumentally enhanced, telescopic optics in early modernity. Trajectories of improvements led to bigger and better telescopes, but these remained limited to optical light. Until the accidental discovery of radio astronomy in the early twentieth century, radio emissions including the background radiation of the universe were unknown. But since radio astronomy is at least analogous to hearing, this is not yet what I am calling translational. Rather, emissions which are "real" but which exceed any of our senses, are the worldly phenomena which can now be translated. Reverting to animal embodiment, certain birds, some mammals, fish, and reptiles apparently can perceive magnetic lines within the magnetic field. This perceptual dimension I do not believe humans have but with the animals who do have it, navigation, orientation, etc. can be embodied. One rather odd discovery has been made recently concerning cattle and such magnetic phenomena—observers of Google satellite imagery noticed that with the exception of locations near power lines, cattle seemed to line up when grazing along magnetic lines. I, myself have frequently confirmed such observations while traveling by train in Europe the last few years. However, there are many familiar instruments which provide translated perception for us: metal detectors for finding coins on a beach, Geiger counters for detecting degrees of radiation, even compasses for navigating are all ordinary examples of such instrumental translation perception. Indeed, astronomically, we can now image radiation from nanoscaled gamma rays to kilometer-length radio waves, thus making frequencies far beyond our normal perceptual capacities mediatedly experiencable. All of this falls within the purview of a postphenomenology.

To this point I have emphasized perceptual embodiment, from ordinary up through connoisseur to instrumentally mediated bodily perception. However, with instrumental translational perception certain different factors come into play. Postphenomenology does indeed pay rigorous attention to embodiment—with Merleau-Ponty, postphenomenology recognizes that human motility, whole body actional embodiment is involved in all human and even intelligent behavior. But with instrumental translational perception whereby dimensions and aspects of phenomena not directly experiencable different factors emerge. Here I return to the early development of my work in the philosophy of technology. Clearly the signature and best known of this work remains my "Phenomenology of Technics." Its first publication in book form was in *Technics and Praxis* (1979), and then expanded and refined in *Technology and the Lifeworld* (1990). In the later form the phenomenology of technics was reprinted in at least seven anthologies and to date has been translated into three languages. This work analyses human-technology-world interrelationships, of which *embodiment* relations constitute an important type. From the beginning I used scientific instruments as one set of examples, such as some referred to above here. But a second set of these interrelations I called *hermeneutic.* As the term suggests, hermeneutics entails a specialized kind of interpretation. Let us take reading as a practice—reading is clearly a perceptual, bodily activity, but what is being perceived are inscriptions, texts, codes, or also visual displays as in science. All of this is still an embodied praxis—but with a difference which here I relate to *isomorphism.* Galileo clearly saw the moon through his telescope (instrumentally enhanced perception) but the isomorphism remains recognizable in spite of and through the previously unseen features of craters, seas, and mountains. But with what I called hermeneutic relations in both books, degrees of isomorphism begin to change. Instruments still refer to a worldly phenomenon, but begin to vary away from perceptual isomorphism. A dial or mercury thermometer refers to how hot or cold something is, but one does not "feel" how hot or cold it may be outside or inside an environment. The same is the case with most instrumental translational perception. My sailboat's compass is referring to magnetic lines, but I "read" rather than feel them as lines. Granted, much design is often directed towards what would be greater perceptual isomorphism. The x-ray part of the spectrum which shows the rapidly rotating pulsar at the center of the Crab Nebula, complete with its radiation jets from the poles, retains a spatial isomorphism for the radiation, whereas a spectrograph visual display would show only a type of "bar code" of spectral color bands. [I object to the common use of these conventions as "false color" since they are assigned for the users purposes. Assigned color or referential color would be better.]

A more ordinary use translational instrument is an oscillograph-like display which transforms speech into visual graph patterns. These instruments are used by various voice teachers to—for one use—teach how to reduce accents. Each sound shows a distinct visual pattern and if the accent speaker can see how the teacher's pattern differs, the accented speaker can learn to change the pattern by changing the sound. This instrumentation is also a good example of what I call science's visual hermeneutics. The habitual pattern in science is to transform all displays into visualizations—here sound patterns into visually graphic patterns. Even Sonar, originally delivered with sounded pings, soon became a visual display.[12] Instrumental translational perceptions, such as those I have referenced here, may be either translations between sensory dimensions (sound to sight or other intra-sensory translations) or translations from non-perceivable dimensions into sensory ones (magnetic lines or X-rays to visualizations.) These more radical translations are also ways in which new "worlds" become available to human experience. One indirect effect on postphenomenology culture has been that by analyzing and appreciating translational perception, postphenomenologists are shown not to be antiscience, but open to new directions of investigation.

To this point, however, the bat question itself has framed my discussion in a dominantly individual-experiential way. Thus whether the above interrelational examples were without or with instrumental mediation, enhancement, translational forms, the reflexive role of experience remained that of a single, embodied experiencer. Yet, from classical phenomenology on, all experience has been understood to be *intersubjective.* That was one major point from Husserl's *Cartesian Meditations,* and is yet another meaning of Merleau-Ponty's "being outside myself in the world." In my terminology, particularly as worked out in *Technology and the Lifeworld* (1990) intersubjectivity, sociality, is already included in *macroperception* and a *cultural hermeneutics.* I argued that such intersubjective experience is already experienced in embodied life. As a very simple example, take eating—we Westerners employ a skilled cutlery practice, knives, spoons, and forks. These technologies all have long histories, today's table knives go back to daggers; spoons, probably the oldest, go back to wooden and bone ones; forks, also from daggers, multiplied tines from one to two or more. And, we teach our very young—baby spoons—to begin to acquire the bodily skills of this use. And this use is embedded in conventional etiquette and coded significations. Setting the table means the fork on the left and knife and spoon on the right with the blade facing the plate. I well remember the slowness with which I learned a French restaurant convention on my first trip there. I could not figure out why waiters were slow to deliver "l'addition" until someone pointed out that I needed to lay my cutlery at an angle on the upper right side of the plate as a signal that I was done. Still using this example, I want to move to another insight of a postphenomenology. There is more than one way to have human-technology eating skills. Most of us who either live in pluralistic contexts or travel, become adept at the Asian counterpart of Western cutlery—chopsticks and spoons. I learned this skill during undergraduate days, once leaving my Kansas farm context with its meat and potatoes plus diet, and learning in those long-ago days to eat chop suey and chow mein, I learned the use of chopsticks. Many decades later on lecture tours of China, I was pleased to hear many times a complement on how well I used chopsticks! I call this a *multistable practice* and from this humble example I now want to turn to the postphenomenological development of multistability.

My own first investigation into multistability was my first edition of *Experimental Phenomenology* (1977) which dealt with multistable visual perceptions. These were what then were usually called "ambiguous drawings," Necker cubes, duck-rabbits, face-vases, and the like. Much of the earlier work on perceptual reversibility arose from Gestalt psychology and its close relationship with early phenomenology. Almost all examples from these early studies were of two- or, at most, three-stability variations. Duck-rabbits and face-vases, old and young woman drawings were two stabilities, Necker cubes rarely, but sometimes were reportedly three (a flat two-dimensional variation was reported, usually interpreted as a result of observer fatigue). My use of rigorous variational analysis, however, showed the whole series of differently positioned Necker cubes and I developed a *five-variant* version for each of the differently positioned cubes. Similarly in several other cases I took what had been previously noted as two-variant forms to nine or ten variants (circles with squiggles). Finally, utilizing what turned out to be simple orientation variations, I transformed the famous duck-rabbit into a four variant duck-rabbit-squid-Martian quadruple. I did not realize in the mid-1970s, that multistability was also beginning to be used in the sciences. I recently discovered that our university president, a physicist, had been working on multistable optics, with an entry in 1976, and then from the mid-1980s a boom has occurred in many sciences. In micro-chemistry, two atom layers of various materials have been found to produce multiple stacking patterns with $5 < 13$ variants; galaxies have finite but many variants; and even infinities have 7. What I understand today is that by a rigorous application of variational theory to various phenomena, unlike early phenomenology which discovered an invariant or "essence" I found that I was discovering multistablities. As it turned out, this phenomenon was especially useful concerning technologies and their cultural adaptations. Thus, later than *Experimental Phenomenology, 1st edition (*1977) I worked out the complex versions of Pacific compared to European navigational systems as multistable in *Technology and the Lifeworld* (1990), and a whole set of instrumental multistabilities in the second edition of *Experimental Phenomenology: Multistabilities* (2012).

Although with multistability we are rather far from the bat problem, there is one more reference back to the bat and Nagel I want to make: my suspicion is that Nagel's insistence that every subjective phenomenon be connected with a single point of view retains a vestigial privacy notion, it clearly is phenomenologically belied by the polymorphism of experience and the multistability of the simple multiculturalism of eating practices to the higher level more complex practices of navigation to the multistability of the material world which contemporary science is beginning to recognize. As Albert Borgmann has so nicely pointed out in *Postphenomenology: Critical Companion to Ihde,* postphenomenology emphasizes multiperspectivalism interrelated to multistability between humans and their world.[13]

Finally, a few words about this volume: the style which has emerged in postphenomenological research has in part followed that of much STS work. This is well illustrated in the chapters here—researchers chose specific topics, many are "empirical turn" topics which related to specific technologies and most of these are contemporary. Thus cell phones, hospital architecture music and aurality, the politics of disasters, postcolonialism, and digital imaging make their appearances here. Broader topics like the origin of the sign, mediation theory, also are present, and at least four authors deal with multistability and others with the body. I am happy to note that what is missing in this collection is any tone of dystopianism, nostalgia, or the other dark tones found so often in early philosophy of technology. This volume is, rather, much more contemporary and future-oriented and exploits the postphenomenological sense and use of variations and possibilities, which are also a dimension of human-technology interrelations.

NOTES

1. Thomas Nagel, "What is it like to be a bat?" *The Philosophical Review* LXXXIII (4, October, 1974) 435–450. p. 1. [Note: pagination on the following endnotes correspond to the print out of the Nagel article from http;//members, aol. com/NeoNoetics/Nagel_bat. Html, not the pagination from its original publication in *The Philosophical Review*.]
2. Don Ihde, "Epistemology Engines," *Nature* Vol. 406, 6 July, 2000, p. 21.
3. Nagel, op. cit, p. 1.
4. Ibid., p. 1.
5. Ibid., p. 2.
6. Ibid., p. 2.
7. Ibid., p. 8.
8. Maurice Merleau-Ponty, *The Phenomenology of Perception,* translated by Donald Landes (London and New York: Routledge Taylor and Francis Group, 2012. Original French publication, 1945), p. 59.
9. Ibid., p. 99.
10. *Smithsonian Magazine,* 23 January 2014, p. 24.
11. *National Geographic Magazine,* "Call of the Bloom," pp. 128–135.
12. Natalie Angier, "Multiple Infinities," *NY Times*, December 31, 2012.
13. Albert Borgmann, "Mediating between Science and Technology, " in *Postphenomenology: A Critical Companion to Ihde*, ed. Evan Selinger (SUNY Press, 2006), pp. 247–256.

REFERENCES

Don Ihde, *Listening and Voice: A Phenomenology of Sound* (Athens: Ohio University Press, 1976).

_______, *Listening and Voice: Phenomenologies of Sound, 2nd Edition* (Albany: SUNY University Press, 2007).

_______, *Technics and Praxis: A Philosophy of Technology* (Dordrecht: Reidl, Publishing, 1979).

_______, *Technology and the Lifeworld: From Garden to Earth* (Bloomington: Indiana University Press, 1990).

_______, *Experimental Phenomenology* (New York: Capricorn Books, 1977).

_______, *Experimental Phenomenology: Multistabilities, 2nd Edition* (Albany: SUNY Press, 20112).

Maurice Merleau Ponty, *Phenomenology of Perceptin*translated by Donald Landes (London and Neew York: Routledge Taylor and Francis Group, 2012).

Introduction

This volume captures a new field in the making. Postphenomenology, an auspicious initiative in technology studies pioneered by the philosopher Don Ihde over the past four decades (Ihde, 1971, 1973, 1976, 1979, 1983, 1986a, 1986b, 1990a, 1990b, 1993a, 1993b, 1998, 2002, 2008, 2009, 2010, 2012). It has attracted ever more attention among a variety of scholars engaged in technology studies, who have found in its concepts and images ways to express insights that, until now, they could not express in a satisfactory way (Selinger 2006). These papers originated at the "Ihdefest," a conference held in Manhattan on March 22–23, 2012, to celebrate the impending retirement of the field's founder.

Three things are distinctive about this collection. First, the nature of the conference at its origin means that the authors of the papers had especially close ties to Ihde, having worked with him in some way or other in the course of his career. Most of these contributors had been Ihde's students, whether as undergraduates, graduates, or postdocs. Nearly all have attended at one semester or another of his famous "Technoscience Seminar" at Stony Brook, which he taught regularly for a decade. All have presented papers in the postphenomenology panels that he arranged at the annual SSS or SHPS conferences, or collaborated with him in some way or other. All of the authors in this collection, that is to say, have benefited from the opportunity to work up close with Ihde.

A second distinctive feature of this collection is the diversity of the commentators: 18 people from 10 different countries, representing the international sweep that postphenomenology has come to have.

Yet a third distinctive feature is that this collection captures a promising young movement in its young condition. This is not just a collection of case studies. Some of the papers seek to analyze, clarify, and develop postphenomenological language and concepts such as "multistability." Other papers expand Ihde's own work, in an area such as music, for instance, into new phenomena in that area that have developed since Ihde's first efforts. Still others are scouting into fields that Ihde had not tackled, such as its political dimensions.

Some papers adopt a reflective approach to postphenomenology, analyzing in what ways postphenomenology has surpassed phenomenology, and in what ways has simply corrected phenomenology and corrected its original promise. Others take a critical tack, and seek to identify postphenomenology's "pressure points," or places where its tools are not yet entirely adequate to the challenge and need more work. Still other papers engage the challenge of postphenomenology itself in an environment where it is likely to be misinterpreted not only as addressing the wrong subjects, but even for being unphilosophical. Postphenomenologists need to anticipate and prepare for such misinterpretations, lest the movement itself suffer.

What is distinctive about this collection, in short, is that it puts on display the diversity of content and styles in a young movement. When a field is older, its research tends to be codified and systematic, its argumentation tends to be standardized, and collections of its research tend to contain pieces that resemble each other in terms of content, methodology, and style. That is not the case here. This collection shows what it is like to be involved in the early stages of a field's development, where the practitioners are drawing on their own experiences to say things they have long felt but only now can articulate.

REFERENCES

Ihde, D. (1971). *Hermeneutic Phenomenology: The Philosophy of Paul Ricoeur*. Evanston: Northwestern University Press.

———. (1973). *Sense and Significance*. New York: Humanities Press.

———. (1976). *Listening and Voice: A Phenomenology of Sound*. Athens, OH: Ohio University Press.

———. (1979). *Technics and Praxis: A Philosophy of Technology*. Dordrecht: Reidel.

———. (1983). *Existential Technics*. Albany: State University of New York Press.

———. (1986a). *Consequence of Phenomenology*. Albany: State University of New York Press.

———. (1986b). *Experimental Phenomenology, second printing*. Albany: SUNY Press.

———. (1990a). *Instrumental Realism: The Interface between Philosophy of Technology and Philosophy of Science*. Bloomington: Indiana University Press.

———. (1990b). *Technology and the Lifeworld*. Bloomington: Indiana University Press.

———. (1993a). *Philosophy of Technology: An Introduction*. Paragon House.

———. (1993b). *Postphenomenology: Essays in the postmodern context*. Evanston, Il: Northwestern University Press.

———. (1998). *Expanding Hermeneutics: Visualism in Science*. Evanston: Northwestern University Press.

———. (2002). *Bodies in Technology*. Electronic Mediations Series. Vol. V. Minneaplois: University of Minnesota Press.

———. (2008). *Ironic Technics*. Automatic Press/VIP.

———. (2009). *Postphenomenology and Technoscience: The Peking University Lectures*. Albany: SUNY Press.

———. (2010). *Heidegger's Technologies*. New York: Fordham.

———. (2012). *Experimental Phenomenology: Multistabilities, second edition*. Albany: SUNY Press.

Selinger, E. (2006). *Postphenomenology: A Critical Companion to Ihde*. Albany: SUNY Press

Chapter 1

Postphenomenology's North American Future

Robert C. Scharff

Does postphenomenology have a future in North America? Of course it does. Don Ihde's Stony Brook beachhead has seen to that. But will it become a deeply influential and widely received movement? That, I think, is still an open question. Since I come to the philosophy of technology from Heidegger, postphenomenologists may hold the doubts I raise here against me. But I am not an old-guard romantic or a dystopian metaphysician—even if today, to coin a phrase, that is what "one says" about those who do philosophy of technoscience in the company of Heidegger. But I shall ignore the *Gerede*. True enough, my remarks about the postphenomenological approach to technology are often critical, and they are in fact often inspired by Heidegger; but they are also positively intended and designed to reduce postphenomenology's vulnerability to another kind of criticism that is not nearly as well-intentioned.

To summarize my general view: On the one hand, I applaud and encourage what has come to be called the "empirical turn" in technology studies. Any philosophically adequate treatment of Western technology in the twenty-first century must start from our actual experiences with technology, must treat these experiences pluralistically—that is, as a multistable, contextualized, and embodied web of material human practices—and must interpret these practices as "*technoscientific*" practices, so that technology is no longer depicted as something generally separate from and mostly subsequent to science.

On the other hand, I disagree with an assumption that appears to be widely held among postphenomenologists—namely, that once we embrace the empirical turn, we can simply get on with technoscientific studies and respond to critics more or less ad hoc, whenever it seems called for. As I have argued elsewhere, assuming this position often expresses a hidden voluntarism—that is, a fairly traditional, ahistorical sort of self-understanding in which one imagines that the way to get over hegemonic and occlusive ideas is by just deciding to do so (Scharff 2010, 109–12). But being human is being historical all the way down; hence thinking of the empirical turn as a choice can be a serious handicap. It has led, for example, to free-wheeling and often inconsistent appropriations of phenomenology, hermeneutics, pragmatism, and post-positivist analytic philosophy, constructed without any sustained analysis of what is actually meant by "empirical," "phenomenological," or "pragmatic," and with an only cursory nod toward the contested nature of these ideas. One does not have to be a Cartesian to want some clarity here. After all, a whole menagerie of philosophers have been empiricists, and pragmatists, and even phenomenologists, so it is not enough for postphenomenologists to reply, "That's not what we mean," whenever they are misunderstood. In what follows, I consider this issue further, only this time starting from some predictable misunderstandings of postphenomenology by others, rather than from problems created by postphenomenology's own self-conceptions.

Stated quickly, my point will be this. Postphenomenologists are entirely justified in claiming that they are on to something that is important, timely, neglected, and often distorted by traditional philosophical methods and assumptions. But their enthusiasm about this discovery often goes hand-in-hand with a tendency to underestimate how deeply uncongenial the dominant philosophical mood in English-speaking countries really is to their discovery. Of course in a general way, everyone knows that North American philosophers are less likely to welcome the interests and projects of phenomenologists, postphenomenologists, pragmatists, or critical social theorists. Yet what often goes unremarked is how mainstream preferences operate more or less *en passant*, in a global, implicit, and unreflective way, not in terms of specifically defended theories or turf conceptions. Hence, even when topics not usually associated with analytic philosophy are recognized, it will very likely be in an inappropriately stylized form.

To put it briefly, no analytic philosopher self-identifies as a logical empiricist today, or follows the old party line about what topics are and are not worthy of attention; but this does not mean that all self-identified postpostivists come genuinely "after" positivism. It is useful to remember that one of logical empiricism's best-known opponents—indeed, the figure that many historians of analytic philosophy now depict as the pioneer postpositivist—is Quine, who famously reassured his fellow revisionists that even after one leaves the dogmas of logical empiricism behind, "Philosophy of science [is still] philosophy enough," and "To be, [is still] to be the value of a variable."[1] In other words, both epistemologically and ontologically speaking, what has displaced logical empiricism is often not really *post*positivism but positivism lite. What has disappeared is a set of theories and doctrines. What lingers is a philosophical atmosphere; and in this silently maintained atmosphere, post-phenomenology will automatically tend to be misjudged—not just for caring about strange topics, or for making odd claims, but because whatever it tries to do will seem in principle and in spirit foreign and unphilosophical.

POSTPHENOMENOLOGY "FRAMED"

To see how this sort of misunderstanding functions, imagine submitting a postphenomenological paper to one of the usual mainstream venues—pick your own conference or journal. The 2009 "Philosophy of Technology" entry in the prestigious online *Stanford Encyclopedia of Philosophy* gives us a pretty good idea of how the paper will be received. (In what follows, my phrasing is close paraphrase of the entry; the tone of voice is my own.) Your paper will first of all quite likely be identified as representing the "humanities philosophy of technology" tradition—not exactly in Carl Mitcham's original sense (although he is cited),[2] but in the sense that your approach is still rather old-school, preoccupied with "experience," maybe also because you mention "historical and social issues," and speak in a jargon that sounds too abstract to be well informed about the latest technological practices. Most importantly, however, the reviewers of your paper will know that your paper is not representative of that

> form of the philosophy of technology [which] has been developing [since the 1960s] that can be regarded as an alternative to the humanities philosophy of technology . . . [and that] has gained momentum in the past 10 or 15 years . . . [so that] it is now becoming the dominant form of philosophy of technology.[3]

This newly "dominant form" of the philosophy of technology is, of course, a branch of analytic philosophy, which has discovered a way to colonize yet another subject matter by discouraging the old preoccupation with history and "the relations between technology and society" and redirect us toward an investigation of the goals, concepts, and methods of "technology itself," with an eye to eventually linking these topics up with the familiar "core issues" of analytic philosophy. Thus, not surprisingly, the rest of the Stanford entry explains how the new colonists of technology are relating it to the Big Three specialties of the philosophical mainstream. By analyzing the "engineering rationality" of quantification and design, they are fashioning it into a respectable topic for *epistemology*. By "defining" technological "objects," with their properties, kinds, and functions, they are domesticating it for post-Kripkean *metaphysics*. And by considering the use of engineering knowledge in terms of issues about moral rules, agency, and responsible choice, they are packaging the subject for Anglophone *ethics*.

Of course, your paper is unlikely to meet these measures of orthodoxy, but your reviewers might follow the Stanford authors and admit that descendants of the "traditional" humanist movement do sometimes seem to discuss ethics—even if they cite all the wrong sources and still treat technology somewhat awkwardly—that is, not as "itself," but as either "a cultural phenomenon that influences our perception of the world," or as "a political phenomenon . . . that embodies institutional power relations between people." Here, on these extra-orthodox fringes, one can expect to find "cultural" thinkers like Ihde, Verbeek, and Borgmann, "political" theorists like Winner, the neo-Marxists, and Feenberg, and "social" theorists like Latour, Callon, and Bijker. Of course, the cultural thinkers are usually obsessed with subjective experience or something called humanism, and the political and social thinkers are too eager to promote their partisan judgments about the technological misuse of power and authority to "engage in *explicit ethical reflection* on technology" (my emphasis). Yet here and there, they do occasionally raise questions about recognizable topics like moral agency. So, in the end, the fate of your paper will probably be determined by whether your reviewers are in a generous mood. You obviously belong to a superseded philosophical movement; you pay too much attention to history, culture, and politics; and you care too much about how we experience technology and too little to what it "is," BUT your paper might somehow be "interesting" in spite of itself.

I exaggerate only a little. The Stanford entry fairly bristles with concepts expressive of a kind of thinking so self-confident in the rightness of its own orientation that for the past 75 years it has often denied that it even is an orientation. As its authors say, they are just reporting what is actually going on in Philosophy—no adjective. For such a Philosophy, postphenomenologists—no matter how young, or technologically savvy, or empirically grounded—are by definition out of touch and mostly out of sight.

Obviously, in the larger world, postphenomenologists are neither. As it happens, I actually enjoyed my first look of the Stanford essay. I had just returned from a round of technoscience studies conferences in Europe and reading this entry made me feel a bit like I had stumbled upon something written in Elizabethan English. All the traditional empiricist-positivist conceptual machinery of mind, intention, sensation, perception, material object, properties, functions, classes, and kinds structures the entry from beginning to end; and its preferred approach is repeatedly described, in familiar postpositivist fashion, as internalist rather than externalist. Fully half the entry's references will be familiar to analytic philosophers who have never heard of technology. Mitcham's idea of humanities philosophy of technology as a historical source of contemporary work is misused to damn all current nonanalytic work with faint praise. And it goes without saying that the entry takes no notice of any of the Continental work that tries, not only to rethink the cultural-political critiques of their forebears but also to "move beyond" the very dualisms of subject/object, inner perception/outer material world, contextless autonomy/external determination that the Stanford entry still takes entirely for granted.[4]

Deconstructing this situation is a complicated affair; but let me underscore one part of the problem that has received too little notice. No matter how loudly they assure us to the contrary, analytic philosophers of technology still understand the whole field of philosophy in a way that frames technology as applied science. In order to recognize this, however, one must look behind what they actually say about science and technology. Here is one of those cases where, as Heidegger puts it, what blocks access to a phenomenon is not explicit theories or concepts, but rather the fact that one thinks about it in an unacknowledged ontological atmosphere—a worldliness, clearing, given and fated site of presencing—in which many phenomena are encountered and theorized about in a less than phenomenological way. In this context—and to focus here only on the modern part of the story—analytic philosophy of technology, like analytic philosophy generally, stands in the shadow of Descartes' tree of knowledge.

According to Descartes' famous metaphor, this tree has three main parts. Metaphysics tends the roots; scientific knowledge of nature ["physics"] constitutes its trunk; and medicine, morals, and mechanics form its three main branches. Descartes' own specific interpretations of metaphysics, physics, and their requisite epistemology were all contested from the start, of course; but there has always been less criticism of his ultimately practical conclusion, namely, that philosophy's principal benefit comes last, in the form of the fruit of its three branches of application.[5] Here, Descartes gives a full and systematic elaboration of Bacon's slogan that knowledge is power (Schouls 1989, 173). The *Meditations* tell us that figure, number, and nature are what can be known; the *Discourse* tells us that we can know these things by becoming objective minds that put aside religion, politics, axiology, common sense, and anything else that is historical or cultural. And the *Principles*'s metaphorical tree depicts the conclusion that technology is just the totality of means for applying our knowledge of figure, number, and nature to do or produce whatever we choose.

Of course, many analytic philosophers would be uncomfortable actually embracing this map of how things are in all of the circumstances in which they nevertheless implicitly understand it to be true. Who, for example, is eager to characterize contemporary medical care as merely applied science, so that surgery is no different in principle from plumbing? Moreover, Descartes' metaphor now operates in more respectable contemporary disguises. His mathematicized metaphysics of the Book of Nature is now epistemological naturalism; medicine has gone off on its own to be the "medical *science*" industry; and his somewhat primitively conceived "mechanics" has matured into "engineering," whose special "rationality" the new philosophers of technology are now busy analogizing to the mathematical and empirical forms of scientific rationality. This leaves practical-minded philosophers with nothing left to do except monitor the "ethical" ideas and behaviors associated with our efforts to achieve the Good Life, which in the "developed" world is routinely depicted as a kind of ever-better product, or planned outcome of scientific and engineering knowledge.

The point is, if you want to be a Real Philosopher, you stay "inside" this atmosphere; breathe elsewhere and you become a mere historian, humanist, or socio-political theorist, with too little interest in mathematics, science, or technology "themselves." Granted, you may sometimes say interesting things, but you can never contribute directly to our understanding of either "knowledge itself"—in this case, engineering or technical knowledge—or the ethical applications of knowledge, since "ethics" includes only the "actions and decisions made by [individual] persons" who employ this knowledge.

POSTPHENOMENOLOGY "[SELF-]DEMOTED"

Postphenomenology's problem, then, is that in the philosophical atmosphere just described, all its efforts—for example, to make a philosophical turn that is "empirical" but not Humean and to redefine "science" as a technologized human practice (Ihde), or to reconceive "objects" as human/non-human hybrids and to take an "ethical" turn in which agency and morality are shared with technical artifacts (Verbeek)—all such efforts will "obviously" be suspect, put on the defensive, guilty of promoting (as one of Verbeek's critics says) philosophical "Newspeak."

Newspeakers do, of course, have good grounds for their turns and redefinitions and hybridizations, but it is also true that the terms they use in their practice are often variations on same terms that define the unwelcoming atmosphere of the tradition they see themselves "moving beyond." Postphenomenology may be a new sort of empiricism, and hybridized moral entities, a new kind of entity, and so on, but to those whose native language is not Newspeak—that is, to those whose philosophizing does not begin with an explicit (and socio-culturally attuned) sense of being experientially in-the-midst of technoscientific life—none of these moves will be intelligible. Instead of listening to Ihde's new empiricists describe technological embodiment and mediation, they will hear strange externalist "claims" being made about how technological things "feel" or are "intersubjectively interpreted." And instead of listening to Verbeek's description of hybrid moral responsibility, they will only hear very "unusual accounts of freedom and agency." Moreover, by employing modified versions of traditional concepts to describe technoscientific life, postphenomenologists themselves risk overestimating the degree to which they actually "move beyond" their predecessors. There is no way to clarify this point without examples. I give two here—Ihde's idea of "expanding" hermeneutics, and Verbeek's idea of "moralizing" artifacts.

IHDE ON "EXPANDING" HERMENEUTICS

Note first that when Ihde expands hermeneutics by viewing science as an *interpretive practice*, he "misses" the epistemic point of Anglophone philosophy of science—which is to study the nature of scientific thinking, not consider scientific activity as a social fact. Indeed, Ihde tells us that even many Europeans who had no problem contextualizing epistemology nevertheless objected to his suggestion that nature's mathematization is just as "interpretive" a practice as historical-human understanding.[6] To vary one of Ihde's images, analytic philosophers continue to picture Galileo without a body or a telescope, but conservative Europeans see no philosophical reason to talk about his use of either one.

Against both objections, Ihde is impatient. Dilthey already shows us that even natural science is an interpretive practice, and he already demonstrated that the focus of hermeneutics thus had to be expanded from merely textual interpretation to cover scientific activity. And given developments of last 100 years, we ourselves see that it is now necessary to expand hermeneutics still further, this time in the direction of technological activity as well.[7] In Ihde's view, with logical empiricism a thing of the past, it has become just obvious that scientific interpretation is not merely some sort of reconstructable cognitive operation.[8] He cites Charles Taylor's famous image of "old-guard Diltheyans, their shoulders hunched from years-long resistance against the encroaching pressure of positivist natural science, suddenly pitch[ing] forward on their faces as all opposition ceases to the reign of universal hermeneutics" (Taylor 1980, 26).

Yet there is trouble brewing here that Ihde does not adequately acknowledge. Rouse, for example, points out that the mere recognition of the new "reign of universal hermeneutics" has not guaranteed full appreciation of its significance. For one thing,[9] analytic and continental philosophers work with two entirely different conceptions of "interpretation," and only one of them—namely, the sort of "hermeneutics of practice" found in Heidegger and the later Wittgenstein—points in the direction of Ihde's expanded hermeneutic-phenomenological study of technoscientific experience.[10] The other conception of interpretation, most closely identified with Quine and Davidson, is a hermeneutics of linguistic translation that still reflects the commitments of the old philosophy of science. For Quine, notes Rouse, it is "a question of my deciding which sentences to accept. For Heidegger, the question is how can I be freed to encounter what is at stake, what is truly questionable, in living now."[11] Hence, although Rouse and Ihde both aspire to be more than postpositivists, their pictures of how to go on from Dilthey are very different. Both want to shift our focus toward the whole complex practice of acquiring knowledge. But Rouse argues that this move requires a detailed defense against a postpositivist tradition that thinks that it, too, is taking the hermeneutical turn. I am on Rouse's side here. Something like Ihde's direction of inquiry deserves to win, but it will not do so in North America without a sustained critical analysis of why it should.

A passing remark by Taylor highlights what is at stake here. For analytic philosophers, he says, the universalization of hermeneutics after Dilthey is a sort of "pleasing fantasy." What he means is that when they gesture toward the idea that all science is interpretive, they do so largely because they are happy with how much progress one seems to make by simply opposing logical empiricism.[12] Postpositivists thus purchase their notion of the universality of hermeneutics too cheaply—by naively over-valuing their opposition to a set of explicit doctrines and thereby continuing to silently embrace the modern understanding of Real Philosophy as one that focuses on the epistemology of natural science-like knowledge and constructs an ethics of its applications.[13] Effective opposition to this (in fact, merely weaker) positivism, argues Rouse, will only come through working out a full-blown alternative conception of science and technology studies that is based on the recognition that "theorizing is as much a practice as any other aspect of scientific work."[14]

In my view, without something like Rouse's systematic diagnosis of what it will take to actually undermine the pre-hermeneutical account of science, Ihde's expansion of hermeneutics remains vulnerable to attack, both externally from analytic philosophers of technology, but also from within the technology studies community, where not everyone is a postphenomenologist. Why, for example, is Ihde's first hermeneutically expansive move to the scientific laboratory, and to technological mediation? Rouse, for one, argues to the contrary that the first order of business should be to stress that science is a *social* practice, and thus to stop treating this practice primarily in Ihde's still "subject-centered" way. Others, like Feenberg for example, also argue that turning too quickly toward a focus on the laboratory and on individual mediations, however phenomenologically well-intended, obscures the sociocultural dimensions of the context in which experiments are pursued and individuals use instruments.[15]

Ihde says little about such objections.[16] Of course, he has no trouble showing how remote traditional accounts are from the realities of scientific vision and instrumentation and how science always involves interpretive practices.[17] But is Latour (or Ihde) more right here than Foucault? As Rouse notes, Latour tends to ignore Foucault's larger argument that the laboratory is only one of the "microworlds" (others being clinics, asylums, schools, factories, and prisons) within which an "intertwining of powers and capacities" is shaped into regulated systems and "extended outward" into the larger world (Rouse 1987, 104–108).

Of course, postphenomenologists are perfectly free to turn to these other "larger" matters in due time, but how should they respond to someone who objects that as a *phenomenological* matter, technoscientific mediations always *presuppose* a culturally complex (Latour), diverse and maybe even undemocratic (Rouse, Feenberg, Foucault) situation, so that turning to these matters "after" discussing embodiment, hermeneutic, and alterity relations always comes too late?

And these objections come from friends! Even a better defense of hermeneutical studies will not displace the powerful, pervasive, and unfriendly tradition still inhabited by analytic philosophy.[18] All the long-standing assumptions that form its ontological default position concerning reality, experience, knowledge, and practice continue to function, in Rouse's colorful image, like "vampires, the philosophical undead that still haunt our concepts and interpretations of nature, culture, and science"—and are still "deeply entrenched in familiar patterns of talk and thought, periodically emerging to vex even those who aspire to surpass them."[19]

The young Heidegger calls this a "pre-philosophical" problem, and thus the initial problem for hermeneutics: He asks, What goes on in philosophy before it becomes what it is?[20] How, for example, does Ihde know to give priority to our technologically mediated foreground rather than the sociocultural background when he starts to write *Technology and the Lifeworld*? I will return to this issue in a moment, after a somewhat briefer consideration of Verbeek's new book.

VERBEEK ON TECHNOLOGY/HUMAN HYBRIDS

In *Moralizing Technology* Verbeek considers technological mediation from the axiological rather than epistemological side of the modern divide and seeks to expand and reconceptualize "ethics" (Verbeek 2011, ch. 4, esp. 87–89). Like Ihde, he argues that both technological mediation itself and our ethical concerns about it must be approached from within our experiences of them—which is what he thinks the modern approaches to ethics have failed to do. What these approaches miss, he says, is that technologies themselves mediate our choices—which mean that it is impossible to give a phenomenologically adequate account of technoscientific life unless we "give up the separation between things and humans" and see that it is really technology-human "hybrids" to which we should attribute agency and responsibility.

In a recent book symposium, Verbeek accuses his critics of resisting his approach because they are still bewitched by what Foucault calls the blackmail of the Enlightenment, according to which one must hold either that our maturity and moral responsibility depend on exercising rational autonomy or that we "[let] things decide for us what to do."[21] Faced with this forced option, any attribution of agency or responsibility to things does indeed seem absurd. But has Verbeek's own approach really avoided the modern metaphysics of subjects and objects? What are we to make of his starting with the very same traditional and "humanist" vocabulary, but then trying to "extend" (how is this done?) the meaning of all its terms?

Verbeek is surely right to reject any philosophy that still takes its modern metaphysics straight. Yet his own discussion often seems to depend heavily on the same metaphysics, only now with a negative sign in front of it. We are urged, for example, to be "*a*modern" and "*post*humanist" and so to include both humans and material things in our ethical considerations because we now see that they are "intertwined" (though also "distinct") (Verbeek 2011, 39–40). We are told that following out this idea of technoscientific intertwining—one in which artifacts and humans both display intentionality, agency, and responsibility but do so "differently"—is the only way to do justice to the moral significance of artifacts. But is it? Verbeek and his symposium critics both start from the legacy of humanism and modernism. Does his transformation of them constitute a genuine "move beyond" them, and if it does, is it the only one? An answer to this question would require reflection on what one must already understand about technoscientific life such that the "intertwining" of humans and material things seems possible and necessary, that it is phenomenologically clear in what sense different sorts of entities display "different" species of agency, etc. But this "pre-philosophical" question is not asked.

Again, my remarks are meant to express friendly misgivings, not "humanist" and "modern" objections of the sort Verbeek faces in the symposium. Following Ihde, Verbeek is surely right to insist that twenty-first century life—at least in our part of the world—is technological all the way down, and that no human activity is fully conceived without recognition of the "materiality" of all our worldly relations. But the myth of moral autonomy will not be put to rest unless one has already made the empirical turn. Hence, if one's critics still understand life in terms of free subjects and external objects—and indeed think they have (in the traditional sense) "empirical" grounds for doing so—it will not be enough to just to assert that moral autonomy is a myth.

The problem, as I have indicated and as Verbeek acknowledges, is both historical and cultural. Modern ontology privileges natural science. Modern culture accepts from this ontology the scientistic dichotomy of factual knowledge and chosen values. But the sticking point is not just the survival of this general legacy. It is the fact that this legacy seems to legitimate putting epistemology and naturalistic metaphysics on the one side, an individualist ethics of agency on the other, and everything else at arm's length. The "obviousness" of this set-up is clearly relied on by Verbeek's symposium critics. As long as one thinks from this obviousness, amodern and anti-humanist claims will continue to be regarded as too remote from Real Philosophy to have any critical power over epistemology, metaphysics, or ethics in their "obvious" forms. From this standpoint, if Verbeek's "extensions" of the modern notions of knowledge and ethics are acceptable at all, it can only be *as extensions*, not as a fundamental critical challenge to what is being extended.

Yet even if one is not an analytic philosopher and actually makes the empirical turn, there is still this question. Is Verbeek right, that the only way to expose the abstract and "ethereal" character of the modern notions of material object and autonomous subject to ascribe some kind of moral agency to *things*—to "distribute [morality] over human and non-human entities"?[22] Will it really clarify how it is to "be" morally responsible (i.e., how this is "disclosed" to us phenomenologically) if we reinterpret our mediated situation so that artifacts "have a share" in this responsibility? Phenomenologically speaking, is it really possible to interpret ourselves as aspects of "hybrids"? Or is the notion of hybridization, of artifacts and humans somehow blended together, perhaps still grounded in the very modernism that both postphenomenology and Actor-Network Theory (ANT) rightly oppose?

My point is that in the end, it will have to be Verbeek who addresses these questions, because his critics see no reason to do so. I agree with him that much of their criticism relies on interpretations of subject, object, freedom, agency, and responsibility that only seem "obvious" because of the vampire-like neo-positivism on which they all depend. Yet it is, after all, Verbeek himself who situates his discussion in the domain of "ethics" *rather than* epistemology and then seeks to "extend" traditional interpretation of its topics. Will he not appear to be engaging in Newspeak, then, for *anyone*—unless they already accept all three of the following, (a) that Oldspeak does not deserve its current hegemony, (b) that there are experienced phenomena that Oldspeak cannot understand, and (c) that it is Verbeek's extensions that do proper justice to these phenomena?[23] In Heideggerian terms, what sort of philosopher does one have to "be" to understand these three things? "How" does one philosophize in order to see that technoscientific mediation in the developed world is best described in amodern, posthumanist, and "extended" ethical terms? And is someone who philosophizes in this way genuinely "phenomenological"? As Verbeek knows, even among those who don't start from Oldspeak, this way of getting to the "ethical" things themselves is not uncontested.

CONCLUSION: HOW TO "BE" A POSTPHENOMENOLOGIST

In short, I believe that clearing a space for postphenomenology in North American is a much more reflectively complicated affair than has so far been acknowledged. In part, it will have to involve sustained rejoinders to an entrenched and unsympathetic philosophical mainstream—rejoinders that spell out in detail why it is currently so easy to misinterpret what postphenomenologists are doing. Yet it will also mean clarifying the relation between postphenomenology and other nonanalytic philosophies of technology that insist on starting instead with socio-political critique, or cultural studies of science, or some aspect of technoscientific life other than Ihde's technological mediation or Verbeek's ethics of hybrids. In short, finally, it will mean asking just how far postphenomenology's own self-conception has really managed to "move beyond," not just this or that traditional conception of subjects, objects, actions, and choices, but beyond the very atmosphere in which even "technological mediation" itself can still sometimes sound a little Cartesian—where "intertwinings" of richer subjects and richer objects somehow replace the "interactions" of Cartesian subjects and objects, as if the latter were not so much ontologically problematic as simply too abstract.

Of course, I obviously do not mean that Ihde and Verbeek themselves should stop what they are doing and start explaining postphenomenology. Nor do I think that postphenomenologists can change the current atmosphere by confronting prominent members of the analytic mainstream and showing them the error of their ways. Yet I do think that a dimension of stronger reflective concern for this issue needs to be made part of postphenomenological practice if it is to be a major player in North American philosophy of technology. It is, after all, always the Newspeakers, the outsiders, the interlopers who must show how the obvious is no such thing. In Vattimo and Zabala's nicely exaggerated phrasing, metaphysics (and epistemology and ethics, too, as traditionally understood) defines the worldview of the winners (Vattimo and Zabala 2011, 37–42).

Moreover, I also worry that just "getting on" with the empirical turn will encourage an approach to technology studies that sells the truly radical implications of a *hermeneutical* philosophy of technology short. Quick expressions of assurance that we are at last sufficiently concrete are no substitute for the sort of critical self-reflection that would allow postphenomenologists to determine more precisely what aspects of pragmatism, empiricism, and postpositivism are phenomenologically valuable; and it would also set them up for dialogue with other outsiders who are not convinced that technology studies should attend first to individual experiences of technological mediation with morally implicated artifacts, rather than, say, the socio-cultural experience of oppression, or depersonalization, or even ontological enframing. To simply begin with a confident reworking of earlier "waves" of philosophies of technology may succeed in determining something of what postphenomenology is not, and in discovering everywhere a fabric of materiality to be investigated in further waves. Yet is that enough to insure that all this materiality will actually *disclose* itself; or are postphenomenologists still responding to partially prepackaged, modern *appearances* when they look to "mediations," place socio-cultural sense in the "background," and build an ethics of techno-human "hybrids"?

NOTES

1. Quine 1976, 151; and Quine 1948, 32 [Quine 1961, 13]. Loud proclamations of post-positivism can, of course, be entirely misleading phenomena (Taylor 1980, 29–30).

2. Mitcham (1994). The authors ignore the fact that Mitcham employs this title to identify the historical origins of the very idea of a philosophy of technology, not to describe a contemporary group or movement.

3. Franssen, Lokhorst, and Poel (2009). See also, Franssen (2009).

4. See, e.g., Verbeek (2011); and Berg Olsen, Selinger, and Riis (2009).

5. Descartes (1985, 186). As Descartes reports, "knowledge of everything useful in life" is what he expected to gain but did not find at La Flèche (*Discourse*, 112–13).

6. They objected still more, of course, to his focus on the technological mediations involved in this practice. Ihde recalls Apel's annoyance at "those [non-transcendental] Americans" during an early International Society for Hermeneutics and Science meeting (Ihde 1998, 40–41, 140).

7. Ihde (1998, 39–42); cf., Ihde and Selinger (2003, 134). In concert with most followers of Husserl, Heidegger, or Gadamer, Ihde assumes that Dilthey's conception of interpretive understanding in the human sciences is at least already proto-phenomenological. See, e.g., Sandmeyer 2009, 31–67.

8. Ihde cites Fuller's remark that between 1977–1988, only one widely discussed book suggested extending the logical positivist program (Ihde 1998, 143n.2), but Fuller stresses that the repudiation of its *program* only partially weakened the hold of its modern, natural science-favoring ontology and epistemology. The "pervasiveness of metaphorical malfeasance in the history of [even] the human sciences" is still evident everywhere (Fuller 2006, 95).

9. Actually, Rouse identifies two problems, the first being Taylor's own distinction between the ontologies of natural and human science (Rouse 1987, 49–51). For Rouse, if all scientific activity is interpretive, then all science is in this respect ontologically the same whether we are engaged with material nature or other people. In fact, he argues, we must stop looking for essential scientific "kinds" altogether, so the truly heterogeneous character of our multiple pursuits of truth can more rightfully be conceived in terms of all the conditions that shape them—including cultural, social, political, ethnic, and instrumental conditions. Rouse (1987, 172–76; also Rouse 1996, 242–57).

10. Ihde (1998, 39–42); Rouse (1987, 49–50).

11. Rouse (1987, 48–49). Quinean hermeneutics reaffirms the traditional empiricist-positivist aim to "interpret what is the case"; the Heideggerian follows the phenomenological alternative and "interprets what[ever] is the matter " Rouse (1987, 48; also 50–80).

12. Old positivist prejudices do not disappear once we realize "that the deductive-nomological model is bankrupt, or that we can't sharply distinguish between observation and theory, or that some (other kind of) understanding is essential to natural science" Taylor (1980, 32); cf., Rouse (1996, 101–121); Scharff (2011, 227–43).

13. Laudan dedicates a whole book to showing that postpositivism is "little more that a hiccup in the history of epistemology"—in the end, nothing but "a particularly self defeating form of neopositivism" whose only real effect on the larger world has been to provide "crucial ammunition for a sustained assault on the cognitive authority of the very sciences that positivists held dear" and to transform positivism itself "into a potent tool for resurrecting the very anti-empirical ideologies that it was invented to banish Laudan (1996, 6, 24–25).

14. Rouse (1996, 101–121, 127–33); and Rouse (2002).

15. Feenberg (2010, 129–46). The empirical turn thus tends to disembody and decontextualize social practice in a way similar to mainstream epistemology's focus on the nature of its reasoning. I ignore here ANT, which has its own reasons for not "starting with the subject."

16. In *Expanding Hermeneutics*, e.g., he simply moves directly into the laboratory—"the place, the site . . . where scientific objects are . . . *made readable*" (Ihde 1998, 50). See also Ihde (2002).

17. Ihde (1991, 98–114); Ihde (1998, 151–83); and Ihde (2009, 45–62).

18. For example, in the very well-regarded recent reference volume (Sarkar and Pfeifer 2006) none of the philosophers mentioned in this article is discussed. Rouse is cited only for an early paper on Kuhn, and even Dilthey is cited only for his (unfortunate) theory of worldviews. The editors explain that their *Encyclopedia* does touch on topics such as actual "experiments by practicing scientists" and even "the social and material context of experiments," but the philosopher-contributors maintain the general focus of the book by emphasizing the "epistemic utility" of experimentation (xvii). The editors assure us that in such an authoritative source as their *Encyclopedia*, "with few exceptions . . . the entries . . . are not concerned with the social role of science . . . [however, since] science is increasingly under ill-motivated attacks in some societies, such as the United States . . . [this] puts philosophers of science, because of their professional expertise, under an obligation to explain science to society, and, where ethically and politically appropriate, to defend the scientific enterprise" (xxiv). So much for any "new model" in philosophy of science. Sarkar and Pfeifer's introduction could have been written in 1960.

19. Rouse (2002, 63, 78). I see the vampire problem not only in postpositivism but also in postphenomenology's notion of technological "mediation." For example, Ihde's "two-tiered" conception of microperceptually embodied "plus" macroperceptually socio-cultural "levels" of experience, by concentrating on individual-technology-world relations, automatically puts everything else in the "background"

to be considered "next"—thus recapitulating the traditional subject-object-and-then-later-context paradigm in Husserlian form (Scharff 2006, 132–39).

20. Heidegger (1999, 46); also, Scharff (2012).

21. Verbeek (2011, 27); Symposium (2012, 626–27). See Foucault (1984, 42).

22. Book Symposium (2012, e.g., van de Poel's objections, 620–26, and Verbeek's reply, 627–28, 629–30).

23. In other words, for any outsider, epistemology and ethics will remain part of real philosophy and consideration of history and culture, part of the mere humanities, until it is shown to them in detail how this distinction is an artifact (no pun intended) of the neopositivist legacy and not a piece of free-floating obviousness.

REFERENCES

Berg Olsen, Jan Kyrre, Stig Andur Pedersen, and Vincent F. Hendricks, eds. (2009). *A Companion to the Philosophy of Technology*. Oxford: Wiley-Blackwell.

Berg Olsen, Jan Kyrre, Evan Selinger, and Søren Riis (2009). *New Waves in Philosophy of Technology*. New York: Palgrave Macmillan.

Book Symposium on Peter-Paul Verbeek's *Moralizing Technology: Understanding and Designing the Morality of Things*. *Philosophy and Technology* 24 (4) (2012): 605–31.

Descartes, René (1984). *The Philosophical Works of Descartes*, vol. 1, John Cottingham, Robert Stoothoff, and Dugald Murdoch, eds. Cambridge: Cambridge University Press.

Feenberg, Andrew (2010). *Between Reason and Experience: Essays in Technology and Modernity*. Cambridge, MA: MIT Press.

Feest, Uljana, ed. (2010). *Historical Perspectives on Erklären and Verstehen*. Dordrecht: Springer.

Fine, Arthur (1996). *The Shaky Game: Einstein, Realism, and the Quantum Theory*, 2nd ed. Chicago: University of Chicago Press.

Foucault, Michel (1984). What is Enlightenment? (trans. C Porter). In Rabinow, Paul (ed.). *The Foucault Reader*. New York: Pantheon, 31–50.

Franssen, Maarten (2009). Analytic Philosophy of Technology. In Berg Olsen 2009, 184–88.

Franssen, Maarten, Gert-Jan Lokhorst, and Ibo van de Poel (2009). Philosophy of Technology. In *The Stanford Encyclopedia of Philosophy* (Winter 2003 ed.), ed. Edward N. Zalta, URL = http://plato.stanford.edu/entries/technology/#Jou. Copyright © 2009; accessed November 12, 2013.

Fuller, Steve (2006). *The Philosophy of Science and Technology Studies*. New York: Routledge.

Heidegger, Martin (2005). *Introduction to Phenomenological Research*, trans. Daniel O. Dahlstrom. Bloomington: Indiana University Press.

———. (2002). Wilhelm Dilthey's Research and the Struggle for a Historical World-view, trans. Charles Bambach. In *Supplements: From the Earliest Essays to 'Being and Time' and Beyond*, ed. John Van Buren. Albany: SUNY Press, 147–76.

———. (1999). *Ontology: The Hermeneutics of Facticity*, trans. John van Buren. Bloomington: Indiana University Press.

Ihde, Don (2009). *Postphenomenology and Technoscience: The Peking University Lectures*. Albany: SUNY Press.

———. (2002). How Could We Ever Believe that Science Is Not Political? *Technology in Society* 24 (1), 179–89.

———. (1998). *Expanding Hermeneutics: Visualism in Science*. Evanston: Northwestern University Press.

———. (1991). *Instrumental Realism: The Interface between Philosophy of Science and Philosophy of Technology*. Bloomington: Indiana University Press.

Ihde, Don, and Evan Selinger, eds. (2003). *Chasing Technoscience: Matrix for Materiality*. Bloomington: Indiana University Press.

Laudan, Larry (1996). *Beyond Positivism and Relativism: Theory, Method, and Evidence*. Boulder, CO: Westview Press.

Mitcham, Carl (1994). *Thinking through Technology: The Path between Engineering and Philosophy*. Chicago: University of Chicago Press.

———. (1990). Three Ways of Being-With Technology. In Gayle L. Ormiston, ed., *From Artifact to Habitat: Studies in the Critical Engagement with Technology*. Bethlehem, PA: Lehigh University Press. pp. 31–59 (Reprinted in Mitcham 1994).

Philosophy of Technology (2009). In *Stanford Encyclopedia of Philosophy*. Accessed November 12, 2013, from http://plato.stanford.edu/entries/technology.

Pitt, Joe (2010). It's Not About Technology. *Philosophy and Technology* 23 (3–4), 445–454.

Quine, W.V. (1976). Mr. Strawson on Logical Theory. In *The Ways of Paradox and Other Essays*, rev. ed. Cambridge: Harvard University Press.

———. (1948). On What There Is. *Review of Metaphysics* 2 (5): 21–38. [Revised version in *From a Logical Point of View*, 2nd ed. New York: Harper and Row, 1961, 1–19].

Rouse, Joseph (2002). Vampires: Social Constructivism, Realism, and Other Philosophical Undead. *History and Theory* 41 (1), 60–78.

———. (1996). *Engaging Science: How to Understand Its Practices Philosophically*. Ithaca: Cornell University Press.

———. (1987). *Knowledge and Power: Toward a Political Philosophy of Science*. Ithaca: Cornell University Press.

Sandmeyer, Bob (2009). *Husserl's Constitutive Phenomenology: Its Prospects and Promise*. New York: Routledge.

Sarkar, Sahotra, and Jessica Pfeifer, eds. (2006). *The Philosophy of Science: An Encyclopedia*. New York: Routledge.

Scharff. Robert C. (2012). Becoming a Philosopher: What Heidegger Learned from Dilthey, 1919–25. *British Journal for the History of Philosophy* 20 (1): 122–42.

———. (2011). Displacing Epistemology: Being in the Midst of Technoscientific Practice. *Foundations of Science* 16 (2–3), 227–43.

———. (2010). Technoscience Studies After Heidegger? Not Yet. *Philosophy Today* 54 (5) [SPEP Supplementary Issue]: 106–14.

———. (2006). Ihde's Albatross: Sticking to a "Phenomenology" of Technoscientific Experience. In Selinger, ed. *Postphenomenology: A Critical Companion to Ihde*. Albany, SUNY Press, 131–44.

Schouls, Peter A. (1989). *Descartes and the Enlightenment*. Kingston: McGill-Queen's University Press.

Taylor, Charles (1980). Understanding in Human Science. *Review of Metaphysics* 34 (1): 25–38.

Vattimo, Gianni, and Santiago Zabala (2011). *Hermeneutic Communism: From Heidegger to Marx*. New York: Columbia University Press.

Verbeek, Peter-Paul (2011). *Moralizing Technology: Understanding and Designing the Morality of Things*. Chicago: University of Chicago Press.

Chapter 2

Beyond Originary Givenness?

Postphenomenology, Digital Imaging and Evidentiary Responsibility

Shannon Vallor

The motto of classical phenomenology, "To the things themselves! (*Zu den sachen selbst*!)," is grammatically normative. It is not a description but an imperative. Of course, in the *Logical Investigations* Edmund Husserl framed it rather more politely: "Let's go back to the things themselves." (*Wir wollen auf den Sachen selbst zuruckgehen*) (Husserl, 2001 [1900/1901]: 168). Still, as suggested by the short version of the motto that became phenomenology's rallying cry, Husserl did not intend this remark to be a mere suggestion, but a new methodological imperative. What motivates his imperative? Is it an epistemic aim, an ethical one, or both? The type of normativity at the core of phenomenological practice is never fully articulated in Husserlian phenomenology. Indeed, the descriptive aims of phenomenology stand in tension with the normative grammar of its motto; the practice of describing what appears, just in the way that it appears, seems to require eschewing any preconceived aim or expectation.

Perhaps no living scholar has done more to carry out the descriptive mission of Husserl's imperative than Don Ihde, whose explorations of the multistable dimensions of sound, vision, and touch have greatly enriched and expanded our understanding of phenomena and how we live in and through them. Yet Ihde has never been afraid to challenge Husserlian orthodoxies; he bears the standard of postphenomenology with pride, and is unabashed in his characterization of Husserlian phenomenology as an "albatross" (Ihde, 2003). If Nietzsche was right that "one repays a teacher badly by remaining only a pupil," then any debt Ihde owed to Husserl has been paid in full, and more besides.

In particular, Ihde's concept of multistability undermines Husserl's original claim in *Ideas* to have founded a descriptive science of static eidetic essences; indeed, Ihde's explorations of human-technology relations have shown us that phenomena appear to us in far more fluid and open-ended ways than Husserl understood. This insight motivates Ihde's rejection of the Husserlian notion of essences as "givens." Ihde's work expresses a consistent and often justified suspicion of fixity, which dovetails well with his alignment toward the descriptive, and away from the normative, pole of phenomenology's meaning. Indeed, Ihde's reluctance to engage in normative philosophizing has come in for much criticism over the years.[1] Much of this criticism is rooted in practical and political concerns about Ihde's disinterest in identifying opportunities for constructive reform of suboptimal human-technology relations.

But the kind of normativity implied in the phenomenological motto is not, at least immediately, of a political kind. For the normative implication of '*zu Sachen selbst*!' suggests a different sense of responsibility, one not to human subjects but to things. It is the responsibility to confront them as given, to meet the phenomena exactly where and how they are, rather than where and how my theoretical constructs and schemas expect them to be in advance. Yet as Ihde's rejection of the "given" suggests, does this sense of implicit epistemic responsibility (and perhaps ethical, too) to the things themselves remain hopelessly wedded to naïve modern notions of reason as a mirror held up to the world, rather than an active participant in the construction of the world's meaning?

To interpret Husserl as imposing this sense of normativity would be uncharitable; the core insight of *Ideas I*, one never abandoned by Husserl thereafter, was that the world always and only appears through the meaning-constituting activity of a subject. The mind, or more properly, the embodied living subject, cannot be understood as a mirror held up to the world, for the world is unthinkable without it. Yet the implication of responsibility to the things themselves cannot be easily evaded. For there is such a notion of responsibility implicit in Husserl, one embedded in his account of evidence (*Evidenz*) and the motivational structure of reason that is called to respond to it.

In the first part of this chapter I argue that Husserl's motivational account of evidentiary responsibility in *Ideas I*, once subjected to an important "hermeneutic correction" by Ihde: (1) provides a rich ground for notions of epistemic and moral normativity, (2) one best explicated in the idea of a virtue of attentive-responsive subjectivity, (3) and that an account of such a virtue encourages and can inform normative evaluations of the very kinds of human-technology relations with which Ihde is concerned.

In the second part of the chapter I illustrate point #3 by showing how a Husserlian account of attentive-responsive subjectivity can help us to evaluate the epistemic and ethical merits of concrete human-technology relations, in ways that respect their multistability and variance while enabling us to protect the normative function of evidential practices. In particular, I will show how evidential uses of digital imaging in medical, military and journalistic practice, for which standards of epistemic and ethical normativity are increasingly critical, can be more usefully evaluated by means of this account than by traditional notions of truth-correspondence or mirroring.

EVIDENTIARY RESPONSIBILITY IN HUSSERL AND IHDE

In Section 136 of *Ideas I*, Husserl tells us that "[A rational position] has its original legitimizing basis in originary givenness" (1982 [1913]: 328). Put simply, this means that the primary rationality of a position-taking, such as a perceptual belief, is dependent upon the "in person" appearance of a thing as perceived. Such a belief is "rationally motivated" by this appearance; the belief is a doxic response that has been invited or called forth in a particular way by the appearance of a physical thing. The point is rather simple to illustrate: if I see a deer ahead of me on the trail, the belief that there is a deer on the road is rationally motivated only by the appearance of a deer as standing before me, given "in person." If I were to have the thought "There's a deer!" simply because the word "deer" came to mind, or because I suddenly remembered a photo of a deer I saw earlier in the day, my belief would not be a "rationally motivated" one, because it lacks the kind of evidence that can properly motivate a perceptual belief.

Husserl defines originary evidence as the unity of a rational belief with the originary givenness that motivates it (1982 [1913]: §137). Thus the term "evidence" in this sense is both relational and normative—it names an appropriate subjective response to an objective presentation. Of course, this unity of rational belief and motivating presentation can and often does come apart. It can happen that I think I see a deer "in person" in front of me but I realize soon thereafter that my perceptual belief was mistaken. The originary evidence that obtained initially has failed to sustain itself in time, leading me to realize "No, I am just seeing things." Alternatively, the coherence of the originarily given sense can break down ("No wait, what is that?") and/or be reconstituted ("Oh, it's actually a dog.") Thus the rational character of a perceptual belief always depends for its validity upon further intuitive fulfillment of specific indications of the appearance, as they present themselves in an open-ended horizon of meaning. The front side of the deer indicates other sides; the indiscernible texture of its fur promises to resolve should I come close; its warm breath hanging in the air promises to evaporate in time.

Evidence is neither an "objective" property of "things" nor a psychological property or artifact of "minds." Evidence is not an external "appendage" tacked onto the perceptual relation but a basic quality of that relation

(Husserl, 1982 [1913]: §145). Because originary evidence is relational, and more specifically, correlative, it would not be proper to say that it legitimizes rational belief. Rather, it entails it. This is part of what it means to say that a perceptual belief is "rationally motivated" by a phenomenon; it is invited or called forth in a certain way by the things themselves. A being who did not "take a position" even when presented with something given originarily would either lack rationality or fail to exercise it at that moment. Even a radical skeptic who believes they are a brain in a vat or a dupe of a Cartesian demon, if they are rational, will take the position "there seems to be an [X] there" upon the presentation of an originarily given [X]. A person who is conscious and capable of sensation, but cannot respond to originary givenness by forming any perceptual beliefs at all, cannot exercise reason; whether due to neurological injury or other pathological cause.

More common is a person who fails, either characteristically or in a particular case, to respond with the kind of beliefs "invited" by a particular presentation. Such a person exercises reason poorly. Imagine, for example, a person who is too distracted by a TV crime show to recognize that her injured dog at the door is not just "making a bunch of noise," as she believes, but is actually crying in great pain. This particular example helps us to see how both epistemic and ethical normativity can be embedded in Husserl's conception of perceptual rationality as well-motivated response. Of course, parallel conceptions of normativity can be found outside the phenomenological tradition in notions of "reasons responsiveness" and moral responsibility (Fischer and Ravizza, 1998). However, the philosophical frameworks upon which most such accounts are built are most often descriptively impoverished and theoretically loaded in ways that a phenomenological account of evidence avoids.

Of course, not all beliefs, not even beliefs about my immediate physical environment, are motivated in this way. I can have indirect, "non-originary" evidence of a thing or state-of-affairs, as when I continue to believe that my book is on the table because I saw it there just a moment ago, or because I asked my friend where he left it and he said "on the table." But on Husserl's view, such beliefs to be rational must ultimately refer back to a relation of "primal reason" or "primal truth," that is, to a relation of originary evidence (1982 [1913]: §139). The legitimacy of judgments lies in this primal source of evidence which, even when un-actualized, must retain a possibility of "being made explicit" in "actual perceptions" (Husserl, 1982 [1913]: §141).

On Husserl's account, rational motivation is a structure involving "lines of dependencies" in which one judgment or valuation brings me to another. If I kick this dusty rock, further things will be originarily given from within the horizon of perception, and I will meet each of them in belief: the cloud of dust that will shoot up from the ground, the brown mark that will appear at the end of my shoe, the rock skipping across the ground, the terrified finches just ahead scattering into the trees. Lived experience is shot through, at every moment, with these dependencies. These "webs of motivation" in any concrete case constitute the moral and intellectual logic of one's embodied situation (Husserl, 1989 [1952]: §232, §236). Here is where we see the normativity embedded in Husserl's account: one who allows her judgments to be directed by factors which "do not emanate from the sense of matters currently functioning" in her embodied situation, that is, one who fails to constitute an attentive and appropriately responsive evidentiary relation to her surrounding world, "is driven irrationally" (1989 [1952]: §233). I will argue that the normative standard here can function for particular acts, and in the domain of intellectual and moral dispositions. That is to say, I can be inattentive and unresponsive to the intellectual and moral logic of my surroundings either momentarily, or as a result of a deficiency of character. Likewise, I can be attentive and responsive to the salient intellectual and moral features of my world in a single moment, or dispositionally. In these passages of Husserl, one finds, if one is looking, a solid anchor for a robust account of intellectual and moral virtue.

For Ihde, however, this approach rests on cracked or missing foundations. For he rejects Husserl's account of "originary givenness," subjecting it to a hermeneutic correction:

> [Husserl] follows what, to my mind, was a mistaken heuristic which confuses issues . . . For phenomenology, intuitions are constituted, not given. Only already constituted intuitions are "given" within an already sedimented context. When taken as "evidence" the evidence is strictly indexical (thus hermeneutic). (Ihde, 1986: 190)

and:

> all "givens" are merely indices for the general work of showing how any particular "given" can become intuited or experienced. (Ihde, 2003: 133)

Ihde claims, then, that evidence cannot be understood as a foundational relation that anchors perceptions to fixed "givens" or "facts." Evidence can only be understood as a "hermeneutic rule" to uncovering the "context relativity" of an experience (Ihde, 1986: 190). Such relativity manifests invariant but multistable structures; evidence does not yield the "correctly" motivated or "true" perception of a situated thing, but a malleable (though not infinitely plastic) architecture of its possible embodiment relations to us. Thus it cannot be said simply on the basis of some particular phenomenon how, or to exactly what, we are called by it to attend or respond. There will be multiple possibilities of appropriate doxic response, and these will not be a finite set that can be guaranteed in advance; they emerge and stabilize only in particular contexts of living perception.

But in any particular context of perception and belief, there are certain responses that will not make sense, that will refuse to stabilize or be sustained in an evidential relation. They will be inadequate to the things themselves as they show themselves. Yes, Ihde's hermeneutic correction of Husserl is warranted. But it does not prevent us from employing Husserl's normative standard of rationally motivated response. Which judgments are "rationally motivated" in any concrete case may be contextually relative, but nothing precludes our making such discriminations through careful phenomenological (or postphenomenological) analysis. Ihde's hermeneutic correction and his account of multistability are fully compatible with such norms, whether we are talking about perceptually driven norms of intellectual or ethical judgment.

Indeed, Ihde has acknowledged that moral perceptions that have destabilized over time due to a loss of historical and cultural context must be abandoned or reconstituted in newly appropriate ways. For example, he describes the need to rethink clearly "antiquated or surpassed" ethical stabilities, such as perceptions of bestiality that formerly entailed, as part of the moral and intellectual logic of the perceived situation, holding pigs criminally responsible for the act (Ihde, 2006: 277). Thus Ihde's hermeneutic correction still leaves room for a conception of epistemic and ethical virtue as a disposition to form stable and well-motivated beliefs in response to our lived situation. His correction simply reminds us that judgments of what is, and of what should be, are contextually relative and cannot be fixed as appropriate in advance. To quote Donna Haraway, "the codes of the world are not still, waiting to be read" (1988, 593). Evidence is a fluid conversation with an active and responsive world, not the uncovering of a passive, fixed tableau.

Moreover, the conception of attentive-responsive perceptual virtue that I am developing here, which I shall call evidentiary responsibility, can help us to evaluate the very sorts of human-technology relations with which Ihde is chiefly concerned, for example, the role of digital imaging technologies in co-constituting human knowledge and perception. Though Ihde himself may remain disinterested in substantive normative critique of such relations, I hope to show that such critique is not only of practical import, but consistent with a phenomenological (or postphenomenological) understanding of perceptual evidence.

DIGITAL IMAGING TECHNOLOGIES AND EVIDENTIARY RESPONSIBILITY

In the last few decades, evidentiary practices have been radically transformed by our growing reliance on digital imaging technologies in scientific, military, medical, jurisprudence, media and mass-market applications. Many new imaging methods are driven by the convergence of techniques and applications from different domains, such as 3D medical imaging founded on 3D modeling techniques used in the film industry, or the surgical application of hyperspectral chemical imaging techniques used in military air reconnaissance. Another key factor is the rapid growth in computing power, speed, and algorithmic sophistication used in digital image rendering.

New digital technologies are used in a variety of ways, including non-evidential contexts such as entertainment and gaming, but their evidential use is pervasive: new imaging technologies are brought to bear in civil and criminal court cases; medical diagnosis and treatment; military planning, reconnaissance and targeting; reporting and "citizen journalism," law enforcement and national security operations; mapping, transportation and navigation; civil engineering and structural safety assessments; mining and salvage; and research across the vast majority of scientific disciplines.

New digital images have some distinctive features as compared to older imaging methods. They are increasingly highly constructed from raw data that must be extensively filtered and processed in order to yield meaningful images, yet rendering and visualization techniques have advanced so that this data can more easily be made to appear directly representational. Thus images that are highly constructed and contain many data assumptions, transformations, additions and deletions can appear to the naïve viewer as a mere "picture" of an object or scene, such as a human brain or the inside of a home. Such improvements in data visualization mean that advanced or cutting-edge imaging techniques can be made increasingly "user-friendly" and meaningful to nonexperts.

A second feature of new digital imaging techniques is that in an increasing number of applications, they can be made available and updated in real time, enabling embodied interaction with the imaged reality. For example, a surgeon can operate by means of a computer-imaged map of the surgical area in such a way that throughout the procedure, her eyes can remain focused on a screen rather than the patient's body. Real-time processing also allows increasingly seamless integration of digital imaging into ordinary perceptual contexts. For example, "Google Glasses" or other wearable augmented reality devices can superimpose digital images upon our natural vision of the world, giving us new information about the objects before our eyes. Further incorporation into organic sensory processes is likely to become common, for example, via electronic contact lenses or digital implants for auditory enhancement.

Such developments raise serious questions about how these new human-technology relations will impact human knowledge, judgment and behavior. In what ways are new neuroimaging techniques changing how medical professionals think about persons with dementia or mental illness? How do these same techniques reshape the thinking of judges, juries, lawyers and defendants concerning matters of criminal responsibility? How is drone imaging transforming how military and police professionals understand their work and the value of the lives they track, target, investigate, serve or protect? Could different imaging techniques affect how scientists, policymakers and laypersons conceive of our impact on the environment? In what ways do digital imaging filters and cameraphone video impact how we view the events captured and shared on social media by "citizen journalists?" In short, how have each of these new imaging practices, and the many more we could add to the list, altered the meaning of evidence as a rationally motivated relation between human subjectivity and our lived environment? These are all obviously normatively loaded questions; they are posed not with an attitude of detached curiosity, but with an active practical concern for the epistemic and ethical value of evidence, and an interest in preserving the integrity of our evidential practices.

Some worries about the increasing reliance of our evidential practices on new digital imaging techniques, while understandable and perhaps common, can be shown to be deeply misguided. These are worries that remain in the sway of a naïve understanding of native or "analog" human perception. For example, some might think that in contrast to natural, unaided perception, all advanced digital imaging techniques are inherently deceptive, and that the highly processed, selected and constructed images they yield are simply nothing more than "fakes" with no legitimate evidential value. On this view, any evidential practice will be corrupted if it relies upon images produced by such techniques.

Yet phenomenology and postphenomenology have shown us that even native human perception is always already highly mediated, selective and constructive. Much like most of those who rely on digital images, we ourselves are unable to "see behind the curtain" of our own sensory processing and discern what was the "raw state" of the neurological "data" of sensation. Indeed, such "data" are no part of our experience at all—they are not "given." Phenomenology, psychology and cognitive science have all given us compelling reasons to throw out the model of native human vision as passively receptive, in a manner comparable to an analog camera. Just as do highly constructed digital images, native human perceptions routinely distort, conceal, assume and project features of the world around us. Thus it would be unreasonable to think that digitally constructed images necessarily "mask the truth" about the world around us, unless we are willing to admit the same for our own native sense perceptions.

However, there is another, more sophisticated and quite legitimate kind of worry we can have about the evidential use of digital imaging practices.

For our native sensory faculties evolved with us, in our lived physical and social environments. They are presumably quite well calibrated to produce appearances that, while mediated and constructed, are more often than not reliably geared to our motivational impulses such that they tend to invite or provoke situationally appropriate responses. For example, assume that your brain is predisposed to select and amplify any auditory nerve signal that is patterned after that which would be produced by the cries of a human infant. The fact that this constructive neurological activity conditions your hearing of an infant's cry as originarily given, as there in person and calling for your attention and response more so than many other equally loud noises in your proximity, is not a deficit or false artifact of your hearing—it is what a human ordinarily should hear in an environment where a human baby is crying. That sound appropriately stands out against the auditory background, because in most human contexts this sound is of profound epistemic and ethical salience. Remember, the percept that stabilizes an evidentiary relation is irreducibly contextual—what matters is not that I hear or see what is neutrally "there," as if sounds and sights are equally meaningful for all perceivers at all times and places. What matters is that I hear or see what is actually epistemically and ethically salient for me in this lived situation.

It can happen, however, that I fail to do so. Recall our example of the woman who fails to hear her injured dog's pained cries as pained cries, or as calling her to attend and respond, because she is engrossed in the sound of her television. Here you have an evidential failure that results from an environmental distractor. But some evidential failures can be produced by artifacts of digital imaging. For digital image processing is not calibrated by tens of thousands of years of lived experience in specific perceptual contexts. It is calibrated by humans, or by computers themselves, according to selection and processing criteria that may or may not result in appropriate attention and response. Digital image artifacts often produce responses not geared to the situation in which the perceiver is actually engaged, or in Husserlian terms, beliefs or actions not rationally motivated by the reality that has been imaged.

It is always an empirical and contextual question whether my evidential relation to a digital image is appropriate. For example: a sophisticated digital hearing aid generates a highly constructed image of the auditory environment. But its evidentiary value is high if it helps me to be more responsive to the motivating content of my actual embodied situation. If wearing the hearing aid allows me to answer more questions posed to me, to frown appropriately at sad stories and laugh at more jokes, then its evidentiary value is high even if the auditory sensations I am receiving are highly processed and selective artifacts. If, however, the device were to introduce tonal artifacts that arbitrarily make voices sound tense or angry to me, and I therefore started to respond to my companions in ways not invited by their actual speech and tone of voice, then the device in this context would cause a serious evidentiary problem, one with both epistemic and ethical import.

This means that when we make normative assessments of digital imaging techniques, we must be careful not to confuse evidentiary value and appropriateness with a naïve and positivistic notion of "data fidelity." We must also avoid reflexive or blanket judgments of the techniques themselves, judgments made irrespective of context. For example, it makes no sense to describe all fMRI imaging uses as inherently "deceptive" and "misleading," or all drone imaging as a "morally flawed" perspective on the world—as if, independently of circumstance, there is only one authentic, privileged perspective from which one's lived environment should be seen. Instead, we should ask which of our digital imaging techniques, in which kinds of lived contexts with which subjects, create a noteworthy risk of disrupted, deficient or otherwise inappropriate motivational responses. To put it more simply, we can ask which particular contexts of digital imaging practice are likely to promote individual acts and virtues of evidentiary responsibility, and which may do the opposite. This is a meaningful question, and one that can very often be investigated and answered. Let us see how, by briefly considering three domains of evidential imaging practice.

DIGITAL IMAGING AND CITIZEN JOURNALISM

In a 2012 CNN op-ed on citizen journalism and digital image filtering apps, titled "Why Instagram photos cheat the viewer," photographer Nick Stern called news photos created with these new apps "fake images" and "unethical," in contrast to those that "bring the emotion of the scene to the viewer"(Stern, 2012). Indeed, the code of ethics of the National Press Photographers Association warns its members not to "manipulate images or add or alter sound in any way that can mislead viewers or misrepresent subjects" (NPPA, 2012). NPPA's ethical ideal for photojournalism is to "Respect the integrity of the photographic moment." Yet as Stern's critics rightly note in the many online comments on his piece, photojournalists have always used different film stocks, lighting conditions and aperture/shutter speed modifications to capture or even to create a certain "mood" or focus. Why are these new digital technologies any different? Isn't this just another example of "techno-moral panic?"

However, Stern's point is that it is not the photographer who actually encountered the newsworthy situation making these judgments, but "an app designer in Palo Alto." Does this make a difference? To a phenomenologist, even a postphenomenologist, it should. The real issue is not, however, whether filter apps distort the "originally given" scene, as if that were a fixed datum. We have adopted Ihde's hermeneutic correction to the naïve concept of a fixed given, in favor of a notion of perceptions (and representations) as multistable, contextually relative and necessarily mediated.

Rather, the normative issue with the journalistic use of digital filter apps such as Instagram, Hipstamatic and PictureShow is whether photos taken with these tools successfully disclose the salient motivational structures of a genuine perceptual situation. That is, does the image disclose the specific contextual relations by which the scene and its subjects actually motivated the feelings, judgments and actions of a human witness? A skilled photojournalist is one who is faithful not to "data" or fixed "givens" but to the epistemic and moral logic of the situation she documents. When journalistic images serve a legitimate evidentiary purpose, they convey something to viewers about the depicted reality by motivating thoughts, feelings and emotions in viewers that make ethical and intellectual sense as responses to that lived situation. Of course, not every witness to an event, nor every viewer of a news photo of it, will have the same response. The range of responses that "make sense" is very broad indeed.

But some of the likely responses produced by some digitally filtered images will make no moral or intellectual sense of the lived situation—they may even violate that sense. For example, among the widely circulated news photos of the London riots in August 2011, several were produced by a "citizen journalist" using a digital photo app that, like many, introduces "vintage" effects that mimic 1960s–1970s film stock. The colors are faded, artificial grain has been added, the focus is dull. As a consequence, the depicted reality—such as burned out cars in the streets near Ealing—evokes very little sense of immediacy or urgent presence. Seeing photographic evidence of violent chaos from decades ago normally motivates a very different set of moral and political responses than does seeing evidence of those things having happened just now. Chaos and violence today may move us to feel shock or horror or anger or excitement; to protest, to initiate reform, start a dialogue, crack down on dissent, or to dig in and prepare for more. Depending on the circumstances, some of these responses will be more prudent and just than others, but all make sense as motivated responses to this kind of lived situation.

Yet none of these responses well up so easily from looking at the vintage-style photos of those burned cars in Ealing. The filter's artifact neutralizes the motivating, visceral elements of violence and presents them as part of a faded past—even if the viewer knows the photo was taken only days before. The affective power of the reality is muted; if the images invite or motivate anything, it is detached reflection, not political action. Nor was this neutralizing effect chosen by the photographer as a way to evoke the salient motivational features of what she witnessed. Instead this effect is accidental—introduced into the image by an app programmed without any thought to the meaning of that lived situation, or any other. The digital filter in this case obscures and violates the multi-stable (but not infinitely malleable) political and moral logic of this embodied reality.

Yet one can find many examples of citizen journalism in which digital photo filters do not disrupt the motivational logic of the situation, and in some cases may even enhance it. For example, one widely shared Instagram photo of a wounded protestor in Tahir Square, Egypt from February 2011 uses a filter which creates a blurred halo around the central figure—a dazed young man being carried through the streets by his panicked friends. In the photo your eye is drawn exactly to where it might be drawn had one been standing there, arguably, where it ought to be drawn. The most morally salient feature of the scene is starkly highlighted in a manner that causes one to draw in a sharp breath. Whether the filter effect was a fortunate accident, or deliberately added later, arguably does not make a difference to the evidential value of the image for us. Long before digital photography, photojournalists relied heavily on happy accidents to highlight the salient features of an image. What counts for evidential purposes is whether the image undermines or strengthens our connection as attentive-responsive viewers to the motivational logic of the lived situation and its subjects. Thus the technology itself is not what is at issue. What matters is whether its use in a particular context conveys how a human witness might have felt and been moved to act at the scene.

DRONE IMAGING IN MILITARY AND CIVILIAN EVIDENTIARY PRACTICE

Digital drone imaging in military and civilian practice has very many evidentiary advantages. Compared to older methods of aerial photography, the improved resolution and lower altitudes permitted by drone cameras allow direct observation of morally and legally salient conduct, for example, planting IED's, firing a weapon, or kidnapping a bank patron. The same features enhance scientific perception—for example, allowing us to view parts of a volcano or ocean trench inhospitable or inaccessible to humans, or to study animal motion and behavior in circumstances less disturbed by a human presence.

Beyond improved resolution, the evidentiary value of these images benefits from drones' extended loitering power, greater mobility and the potential for vastly expanded area coverage due to the ability to digitally compose thousands of images into one. For example, the ability to loiter for minutes, hours or days and investigate something via different angles may enable a drone operator to recognize initially suspicious behavior as benign, or vice versa.

In the US military context, an oft-lamented fact about drone imaging is that in current practice, human drone pilots in Nevada continue to suffer PTSD and other stress disorders at roughly the same rate as their peers deployed overseas (Otto and Webber, 2013). This is largely because drone loitering allows pilots to linger on the scene of a strike and verify a target's destruction (or in what is notoriously known as a "double-tap strike," to attack those who rush in to help). Thus drone pilots in many cases are witnessing more of war's horrors than are fighter pilots who drop their bombs and speed off. Drone pilots may actually see the bodies of children being carried out of the smoking building, or the man whose legs have been blown off writhing in the rubble. We must note that however damaging such visions, this underscores one legitimate evidentiary power of drone imaging. For aren't the consequences of our military actions, even those we believe were justified by the evidence, of their own evidentiary value? Aren't they an essential element of the lived situation and its meaning? Should pilots and commanders really be shielded from them? Or are they crucial evidence of whether our total response patterns are proportional, reasonable, discriminating, humane, necessary; that is to say, just?

Do drone images, in some contexts, thus preserve and transmit more of the motivational logic of war than does, say, satellite imaging, or the native vision of traditional pilots? This is an important question, as we can imagine that some military bodies may already be seeking new drone imaging technologies that use special filters and frames designed to denature emotionally provocative content, stripping out more of their stress or guilt-motivating structures. This would not be hard to do; for despite their potential to produce psychological stress in soldiers, drone images arguably already denature the motivational structure of their subjects in many ethically and epistemically problematic ways. Reflect for a moment on the US Defense department's informal adoption of the term "bugsplat" in estimating drone strike collateral damage; is such a dehumanizing vocabulary made more or less likely by a form of imaging that represents a school torn apart by a missile from the same perspective by which I notice the crushed insect under my foot?

Drone imaging has motivational effects that differ markedly from other kinds of military vision, because drone images implicate very different embodiment relations been "us" and "them." The birds-eye perspective of a drone creates not only a spatial distancing between me and my enemy, but the illusory perceptual sense of a natural power differential, of incomparable degrees of freedom and security, of incommensurable modes of being in the world. The subject of a drone camera is not embodied like me; he is a thing trapped in two dimensions, moving on the ground beneath me, to be tracked, studied, measured, but never met or confronted. No degree of image resolution will change that. A human in a drone feed is not pictured as occupying lived space. His or her bodily orientation in the world is not one that a drone operator intuitively recognizes as like his own, human being as such. Nor is it like the bird-orientation into which the drone operator projects himself. The target in a drone image has neither the body of a bird or a human. He is a flat, compressed shadow moving on the earth, to be let free to wander or squashed, but never to be met in my own embodied world.

It is worth noting that drone operation has been associated with a "video game" mentality, in which the very different motivational structures of gaming and war imagery fuse (Singer, 2009). Prior to the invention of virtual reality environments, killing a game character (say, in the role-playing game Dungeons and Dragons) would, in anyone that was not a sociopath, provoke very different feelings, thoughts and beliefs than would killing a human person. This is because the moral and perceptual realities of these two kinds of "killing" differ in ways that are profoundly salient. Yet over time, these differences can be occluded by digital imaging techniques that deliberately frame both realities in similar tones, resolutions, scales and audiovisual effects. Digital rendering techniques used in virtual reality gaming have been applied in military simulators for years; but what happens to the motivational logic of a soldier's environment when those same gaming techniques are used to render the images of his or her real-time conflict reality?

Perceptual and moral desensitization is, of course, seen as desirable from a military perspective when dehumanizing the enemy appears critical to mission effectiveness; but such weakening of the motivational structures of perception is prima facie undesirable from both an ethical and epistemic point of view. For it impedes rationally motivated responses to the lived environment one's actions are engaging. Evidentiary responsibility is a reliable disposition to attend and respond appropriately to the epistemically and ethically salient features of those lived surroundings presented for my perceptual consideration, whether through native vision or through technologically mediated and constructed digital images. Drone imaging needs to be evaluated not simply instrumentally, but also in virtue of its effects on our intellectual and moral character as judges of evidence.

Though I have focused on military applications of drone imaging, the same considerations may apply in many civilian/scientific contexts. A drone that tracks the movements of fleeing refugees, or the migration of endangered elephants, may be of great instrumental value while simultaneously producing images with impoverished motivational structures. A drone image that impedes my ability to see refugees or elephants as creatures embodied like me, acting and suffering in the very same lived world in which I act and suffer, exacerbates an epistemic and ethical flaw in my character no matter what other use it may serve. All else being equal, an evidentiary practice will almost always be better served by images that allow a sense of embodied co-presence in a shared world than by images that make that perspective impossible to adopt.

Yet just as in the case of digital photojournalism, the problem in these cases is not digital imaging technology as such. The problem is its frequent presence in human-technology relations that foster suboptimal evidentiary practices. More epistemically and ethically reliable military and civilian imaging practices are thinkable, and they do not require that we relinquish our digital technologies. For example, we could design digital imaging applications to allow a fuller range of perceptual verifications, for example, combining or alternating aerial and ground level images of subjects. We could introduce more multimodal imaging—for example, syncing audio and visual feeds so that imaged situations are both seen and heard. We could introduce haptic sensory elements into images that encourage a sense of embodied co-presence rather than embodied surveillance from above.

We must be careful, however, to recognize that like all virtues, evidentiary responsibility represents a contextually sensitive and intelligent mean between extremes. Not all lived environments and contexts should be approached with evidentiary interest—not everything that can be seen should be seen. For example, one might foolishly try to "solve" the motivational problem in aerial drone imaging by proposing the proliferation of microdrones that can be deployed in swarms to record virtually any human situation from any spatial perspective, not just from above. Such a proposal would invite serious concerns related to privacy and panoptic control that are unlikely to be justified by the evidentiary benefits. Evidentiary responsibility is a disposition to uphold and protect the integrity of evidential practices, which play an important but limited role in the good life; it is not a disposition to view the world as nothing but a repository of information to be mined.

MEDICAL IMAGING AND EVIDENTIARY RESPONSIBILITY

Perhaps nowhere are advances in digital imaging more significant than in the field of medicine. Medical imaging with techniques such as functional magnetic resonance imaging (fMRI), near-infrared spectroscopy (NIRS), hyperspectral imaging, high-contrast/high-resolution ultrasound and molecular imaging are among the increasingly sophisticated digital imaging tools used for medical diagnosis and research.

The impact of these techniques on evidential practice is profound. For example, the use of false color in fMRI imaging can give such images an inflated rhetorical power, especially for lay viewers. If a neuroscientist testifying in court points to an area of the brain image highlighted in red and asserts that this represents anger, or fear, or deceptive intent, a jury may not understand that the false color is not a picture of the psychological state named, but merely a human interpretation overlaid on an image that actually depicts indicators of blood flow, rather than emotional states. Overreliance on neuroimaging has been cited as a significant risk of their evidentiary use in legal settings; juries and judges may ignore other compelling evidence of brain damage because the damage cannot be "seen" on a scan, or they may overinterpret the evidentiary weight of abnormalities that do "show up" in a brain image.[2] In general, laypersons can be moved by neuroimages to respond to digital artifacts of scientific interpretation as if these were instead psychophysical realities presented originarily, "in person," in the matter of a perceived object.

This is a factor even in older modes of medical imaging. As Peter-Paul Verbeek has noted, ultrasound images present a highly mediated and constructed image of a fetus, enlarged for visibility and framed in such a way that it appears to "float free" as an independent subject rather than an embedded presence in the mother's womb (Verbeek, 2008). Again, the viewer of such an image is not primed to think about the enlargement or the framing; they are primed to respond to a reality as simply "given," as there "in person" as if seen through a window. Such a presentation is implicitly moralized; it motivates certain human responses and not others. The epistemic and ethical worry is that the primary motivators are artifacts of the constructed image rather than the lived reality being imaged. For example, the enlarged image of a seemingly free-floating fetus invites me to respond to it as something that I might hold and caress, something that could already fit into the embodied situation of a newborn infant. This is an illusion fraught with moral, epistemic and medical implications.

Thus medical imaging risks distorting the attentive and responsive dispositions of viewers through a curious reversal of Haraway's "God Trick" (1988, 582). Rather than inducing in the viewer the illusion that she is seeing a given reality from on high, "from nowhere," a viewpoint divorced from living human perspectives, scientific images can also induce the illusion of a different embodied perspective on the subject, one that is not actually a lived possibility. This is arguably just as epistemically and ethically perilous, as it may trigger motivational responses that do not make sense, that is, responses that are not in fact invited by the lived situation.

One way of thinking about this is suggested by Annamaria Carusi and Aud Sissel Hoel, who recommend discarding the naïve ontology of imaged subjects and viewers as independent and "pregiven" realities, and instead seeing these as "intertwined" relativities (2014, 207).[3] The point here is that our evidential use of constructed images cannot be evaluated simply by thinking about the quality of the depiction "in itself." We have to think about the specific motivational relations these uses instantiate; how such depictions engage a particular viewer or viewers in a given context, and how we in turn engage them.

We must therefore distinguish the passive and naïve use of neuroimages as fixed givens from their active use in more sophisticated evidential practices. In the former scenario the rhetorical power of such images may inhibit investigation or critical response. But in the latter cases, in the context of a well-cultivated professional practice, users of these images are expected and actively encouraged to interrogate them (Carusi and Hoel, 2015:162). In such cases highly constructed scientific images rationally motivate rather than derail the researcher's efforts to orient and facilitate a responsive embodied engagement with the imaged subject. Thus when used in cultivated practices of active vision, such images may help to render more explicitly, rather than to conceal, the webs of motivation that constitute the logic of the embodied situation shared by the researcher/physician and the imaged reality.

CONCLUSIONS

Twenty-first century digital imaging technologies do not produce intrinsically "fake images" or distortions of the "given" perceptual situation. There are no fixed givens—all images are mediated, and their evidential import is always contextually relative, even when perceived by the unaided senses. Yet in particular contexts imaging technologies can still pose a significant normative danger, by indiscernibly introducing arbitrary, premanufactured and motivationally distorting links in the evidentiary web that binds a perceiver to his or her lifeworld.

Thus even with Ihde's hermeneutic correction to the concept of "originary givenness," Husserl's analysis of the motivational structure of evidence retains considerable normative power. The phenomenological import of evidence is not what it passively shows as given, but what it actively motivates or calls forth as an intellectual-ethical response to an embodied situation. Thus digital images become normatively problematic only when they inhibit, disrupt or distort the motivation of such responses.

The use of new imaging technologies for evidentiary purposes must therefore be subjected to rigorous and contextually sensitive phenomenological analysis, in order to discern those conditions under which the imaging may disturb the integrity of the perceptual situation's motivational structure, which alone supplies the legitimizing power of perception.

NOTES

1. See Selinger (2006) and (2008) for an overview of such criticisms.
2. On this point, see chapter 8 of Perlin (2013); also Blume and Paavola (2011).

3. Carusi and Hoel's approach works well with mine insofar as they draw from Merleau-Ponty's late ontology of "intertwining," which extends the Husserlian view of perceptions as relational "webs of motivation."

REFERENCES

Blume, John H. and Emily C. Paavola (2011). Life, Death and Neuroimaging: The Advantages and Disadvantages of the Defense's Use of Neuroimages in Capital Cases. *Mercer Law Review* 62: 909–931.

Carusi, Annamaria and Aud Sissel Hoel (2015). Brains, Windows and Coordinate Systems. In *Visualization in the Age of Computerization*, eds. Annamaria Carusi, Aud Sissel Hoel, Timothy Webmoor and Steve Woolgar. New York: Routledge, 145–169.

______. (2014). Toward a New Ontology of Scientific Vision. In *Representation in Scientific Practice Revisited*, eds. Catelijne Coopmans, Janet Vertesi, Michael E. Lynch and Steve Woolgar. Cambridge, MA: MIT Press, 201–221.

Fischer, John M. and Mark Ravizza (1998). *Responsibility and Control: A Theory of Moral Responsibility*. Cambridge UK: Cambridge University Press.

Haraway, Donna (1988). Situated Knowledges: The Science Question in Feminism and the Privilege of Partial Perspective. *Feminist Studies* 14(3): 575–599.

Husserl, Edmund (2001[1900/1901]). *Logical Investigations*, 2nd Ed., ed. Dermot Moran. New York: Routledge.

______. (1989 [1952]). *Ideas Pertaining to a Pure Phenomenology and to a Phenomenological Philosophy, Second Book*. Trans. R. Rojcewicz and A. Schuwer. Dordrecht: Kluwer.

______. (1982 [1913]). *Ideas Pertaining to a Pure Phenomenology and to a Phenomenological Philosophy, First Book*. Trans. F. Kersten. Dordrecht: Kluwer.

Ihde, Don (2006). Forty Years in the Wilderness. In *Postphenomenology: A Critical Companion to Ihde*, ed. Evan Selinger. Albany: SUNY Press.

______. (2003). If Phenomenology is an Albatross, is Post-Phenomenology Possible? In *Chasing Technoscience*, eds. Don Ihde and Evan Selinger. Bloomington, IN: Indiana University Press, 131–146.

______. (1986). *Consequences of Phenomenology*. Albany: SUNY Press.

National Press Photographers Association (2012). *NPPA Code of Ethics*. https://nppa.org/code_of_ethics.

Otto, Jean and Bryant Webber (2013). Mental Health Diagnoses and Counseling Among Pilots of Remotely Piloted Aircraft in the United States Air Force. *Medical Surveillance Monthly Report* 20(3): 3–8.

Perlin, Michael (2013). *Mental Disability and the Death Penalty: The Shame of the States*. Lanham, MD: Rowman and Littlefield.

Selinger, Evan (2008). Normative Judgment and Technoscience: Nudging Ihde, Again. *Techne: Research in Philosophy and Technology* 12(2): 120–125.

Selinger, Evan, ed. (2006). *Postphenomenology: A Critical Companion to Ihde*. Albany: SUNY Press.

Singer, Peter W. (2009). *Wired for War: The Robotics Revolution and Conflict in the 21st Century*. New York: Penguin.

Stern, Nick (2012). Why Instagram Photos Cheat the Viewer. *CNN*, February 22, 2012. www.cnn.com/2012/02/22/opinion/phones-instagram-apps-stern/.

Verbeek, Peter-Paul (2008). Obstetric Ultrasound and the Technological Mediation of Morality: A Postphenomenological Analysis. *Human Studies* 31: 11–26.

Chapter 3

Historical Variations and the Cellular Age

Galit Wellner

What do the tale of Hansel and Gretel and an alarm clock have in common? One answer can be that they both belong to historical periods in which we managed without a cellphone. We used to wake up to the cry of the rooster and later to the buzz of the clock, and mark our way by signs like the white pebbles that were used by the siblings in the story by Brothers Grimm. An eye opener showing how an old technology like an alarm clock has been influenced by cellphones occurred in autumn 2010 at the time of year when many countries switch from daylight saving time to standard time. In that autumn, many Blackberry users were late to work due to a bug in their smartphone's operating system related to the switching of time. Because their smartphones were tuned to a wrong time, they woke up one hour late.[1] This incident illustrates that the cellphone is used not only to contact others (as its name may imply) but also as a helper and a companion that assists people in their timely wake-up. With more recent developments, the cellphone is also used for moving in the world. People use their handsets as navigation devices (instead of relying on pebbles and road signs) even on the well-known way to work. They do so because the cellphone provides more than just the shortest path but also alerts on traffic jams, police "traps" and other road conditions. The alarm clock and the navigator examples reveal how the cellphone surpasses the communication paradigm; it becomes more than a mere telephone; it occupies an increasingly significant role in our lives. Its central role in our everydayness "colors" and determines the age we live in.

That is why I contend that we live in the cellular age. This age is an "offspring" of the Radio Age and the Television Age. Each such age is characterized by a feeling of how-could-we-live-without-this-technology. Naming ages after dominant technologies is based on the belief that each technology produces a different set of habits, moods, experiences, perceptions, manners, modes of thinking, knowledges, all grouped in what Martin Heidegger terms being-in-the-world (*In-der-Welt-sein*). In his book *Being and Time* (1996), Heidegger gives an example of a hammer to demonstrate how everyday tools shape our being-in-the-world. Everyday tools, he claims, constitute the way we conduct ourselves in our environment.[2] Everyday tools are diverse and they differ by the age to which they belong; there is a huge gap between the hammer and the cellphone. According to Don Ihde, "overall, today's technologies evidence a quite different flavor from what was prominent during Heidegger's lifetime" (Ihde 2010, 5). This difference in flavor is exemplified in the shift from hand-operated everyday tools like a hammer (as well as mechanical like cars and hydroelectric dams) to digital everyday tools like a cellphone. There is a different being-in-the-world when one walks down the street with a hammer in the hand than walking the same street with a cellphone in the hand.

This chapter deals with the question of how to characterize the age that we live in by the dominant technologies we use. Before answering this question I will elaborate on technological ages and generations and relate this type of analysis back to Karl Marx and Friedrich Engels. Then I will touch upon Michel Foucault's "history of the present" in order to construct my own permutation—the "histories of the future." After these two preparatory steps, I will discuss the postphenomenological methodology of historical variations and will apply it to cellular technologies. I will provide an historical-genealogical account of the cellular age and the developments it went through; this account is based on the postphenomenological methodology of historical variations that articulates the transformations in technologies, being-in-the-world and the environment into periods differentiated by their style. In the Cellular Age, these periods are extremely short, of no more than several years. And the last part will consist of a comparison between Foucault's genealogy and the postphenomenological historical variations.

THE CELLULAR AGE

What is the Cellular Age? How can a relatively new technology such as the cellphone define an age? Can an age last no more than a few decades? It might be difficult at first sight to consider a Cellular Age in comparison to the Stone Age that continued for thousands of years and whose technologies—like the hand axe—were relatively stable over several human generations.

The cellphone and its fast evolution urge us to rethink what is a technological age. The answer is that a technological age is more than the technology which dominates that period in time; it is characterized also by distinct modes of thought and social practices. This is basically the argument of Karl Marx and Friedrich Engels in *The German Ideology* (1970). In this work, they correlate between Middle Ages' workshop practices and the feudal social structure, and then between early Modernity's factories and the then-emerging capitalistic social structure. Once the capitalistic order took place, the older feudal system seemed to be impractical and even illogical. Similarly, once we get used to compact cellphones that contain our phone books, calendars and other personal information, the previous age of bulky fixed phones looks outdated, limited and unworkable. Heidegger explains this scheme from a technological point of view. He asserts that an ancient silver chalice reveals a different world than that which is revealed by modern technologies such as a hydroelectric dam (Heidegger 1977; cf. Ihde 2010). Each technological artifact belongs to a "network" of things and objects that establishes a context in which the artifact is used and which produces a specific worldview (Heidegger 1996). The worldview changes over time as the context in which the artifact operates changes. This is a process of transformations. Different technologies produce different historical periods, what I call here Ages. In different Ages, the dominant technologies convey different worldviews (and vice versa). Eventually different being-in-the-worlds develop, as Neil Postman writes: "to a man with a hammer, everything looks like a nail . . . To a man with a computer, everything looks like data" (Postman, 1992, 14). An age can therefore be defined by a dominant technology that shapes people's being-in-the-world.

Yet, today's technologies evidence a different flavor. Nowadays a generation of humans experiences in a lifetime more than a single generation of technologies. Moreover, a human generation can experience many technological generations that together comprise an Age. What is different in contemporary Ages is that each of their technological generations can last no more than very few years, as Moore's law illustrates. Gordon Moore, the founder of Intel, observed back in 1965 that every eighteen months the number of transistors on an integrated circuit is doubled. Paul Ceruzzi, historian of technology and curator at the Smithsonian expounds: "the steady and unstoppable march of semiconductor density . . . has led to the rapid introduction of an enormous number of new products, services, and ways of working and living" (2005, 587). With faster introduction of new technologies, the Cellular Age is much shorter than the Stone Age, the Bronze Age, and even the Radio and Television Ages, as it consists of shorter technological generations. Each generation has its own devices through which it confers a slightly different meaning to the world.

HISTORIES OF THE PRESENT, HISTORIES OF THE FUTURE

Marx and Engels place the beginning point of their historical-genealogical analysis in prehistory. When should the story of the Cellular Age start?

Should we begin in prehistory with the practice of smoke communication between two tribes? Or should we begin in modernity, with innovations such as Marconi's radio communication and Bell's telephone? One way to decide on the starting point is to look at the end point, that is—the present. It means that the current understandings are the starting point for the analysis so that we go back in time to explore the past in order to better understand the present. This is what Michel Foucault calls "history of the present" (Foucault 1995; cf. Dreyfus and Rabinow 1982).

But in an age ruled by Moore's Law—when change is frequent and generations are much shorter—the present does not provide a stable starting point. In such an age, I claim, it might be more fruitful to talk about "histories of the future." The Cellular Age's history should be a history of the future, because it is the possible future developments that direct us in reading the history of the cellphone and guide us in extracting from such history the lessons required for possible evolutionary paths. This is the meaning of what I shall call here "histories of the future."

Histories of the future can be positioned within the Heideggerian temporality that rejects a linear conception of time and instead ties present and future so that "the future . . . makes present in the process of having been" (Heidegger 1996, 300 [326]). It is the future that produces the present. The present is not merely positioned between a past and a future, but it is more a "comportment" toward the future. In this process, Ihde explains, "a projected possible future gathers a particular perspective upon the past into a then finally constituted present" (Ihde 1998, 81). The past is weaved into the future as a source of perspectives, so that our understanding of the past reflects our present position and the "comportment" toward the future. There is no single future that is waiting for us; that would be a too deterministic approach. Yet, histories of the future form a way of thinking about the future in terms of opening new horizons thereby revealing the future as a liquid set of changing options.

While attempting to describe and qualify such "a projected possible future" for the cellphone, my postphenomenological work is not engaged in what Ihde calls "technological dreams" or "technomyths" (Ihde 1998, 105). Instead, it is a search for the roots of these future possibilities in the very recent past, refraining from science fiction and imaginary scenarios. Therefore, the methodology of histories of the future can participate in and contribute to the research, engineering and design of everyday technologies. It is part of postphenomenology which is a type of philosophy of technology that offers a practical perspective on technologies. The combined histories of the future and postphenomenology can enrich Research and Development (R&D) processes by unfolding new horizons and future possibilities.

POSTPHENOMENOLOGY'S HISTORICAL VARIATIONS

Postphenomenology offers a methodology to investigate technological developments and the related changes they bring. It consists of analyzing technologies and our relations to them, by "dividing [into] periods by the style of technologies used" while simultaneously reviewing the "practices that correlate with the technologies" (Ihde 2009, 46). These periods are termed historical variations.[3] Each historical variation is designated by a technological style and its related practices. The researcher's task is to extract from multiple practices and functions the prevailing usage pattern which stands for the technological style.[4] This is how a historical variation is identified. Ihde demonstrates the application of historical variations in his analysis of astronomy's practices. He describes the move from the ancient observer who uses "eyeball;" through the modern observer assisted by devices such as a telescope that augments the human vision; to today's observer who deploys extra-optical sensors which are "visualizing the invisible" (Ihde 2009, 45). While Ihde uses the historical variations to articulate a field of scientific knowledge such as astronomy over the ages, I use it on the much shorter historical span of the Cellular Age and for analyzing everyday technologies. In my analysis, each historical variation of the cellphone is a technological generation, and not an age of its own. Yet, this difference is of magnitude, not of essence.

Historical variations of everyday technologies (like the cellphone) can be extracted from sociological and historical research, together with handset manuals, journalistic articles, blog posts and the like. These sources of data serve to identify the dominant discourse as per a technological generation. The discourse reflects the perceptions regarding the cellphone at a given time, that is—the being-in-the-world shared by many people around the world using this technology. In postphenomenological terms, the focus is on the experiences co-shaped by technologies and users. In addition to the historical variations, Postphenomenology also supplies a tool kit for classifying the different relations between cellphones, their users and the environment in the formula of "I-technology-world." The formula reminds us that people, technologies and environments transform and co-constitute each other in a complex process that is not necessarily dialectic or synchronized. Each historical variation should refer to these three aspects, as will be demonstrated for the cellphone in the next section.

CELLPHONE'S HISTORICAL VARIATIONS

Some analyses of the cellphone's history have already been performed. Probably the most common is the engineering account according to which the cellphone's evolution is divided into first generation—mostly analogue, second generation (marked as 2G)—digital,[5] third generation (3G)—optimized to transmitting data, and so forth. This partitioning is taught by industry organizations such as GSMA (Groupe Speciale Mobile Association) and the United Nations' ITU (International Telecommunication Union), as well as mobile operators and technology manufacturers. It is based on the different frequencies and the speeds of the transmission between the handset and the antenna. Sociologists and media practitioners usually examine a certain generation or a certain technological feature of it. Those who study voice communication usually focus on the first generation[6]; those who study text messaging usually focus on the second generation.[7] They base their analysis on the engineering partitioning. Few scholars aim at examining the historical evolution of the cellphone, but they also take the engineering classification as a basis and do not attempt to rethink it. [8]

Practicing postphenomenology, my articulation is slightly different. It derives from the experience users have with their technologies, and not from the measurement of transmission speeds. I examine the relations between the cellphone, the user and the world in accordance with the postphenomenological formula of I-technology-world. The analysis aims to identify the technological styles across technological generations, that is—the ways the handset is used, the associated practices, the applications offered by various manufacturers and the changes in the cultural environment. Using this methodology, I have mapped four historical variations in the evolution of the cellphone: the first is about voice communications; the second—short alphabetic texts; the third—multimedia applications; and the fourth—sensing the environment and deciphering it. The last historical variation is indebted to the "history of the future," though some of its features can be experienced these days.

The first historical variation is about voice communications. It can be roughly identified as the analog cellular technologies of "1G" that are characterized by bulky handsets in a configuration known as "brick."[9] One of the later forms is the "clamshell" configuration which first appeared on January 1996 in Motorola's StarTAC, the last major analog cellphone. Thanks to its small size and low weight, StarTAC was carried in the pocket or on a "holster." Because it was held near the body, Motorola could add vibration as an indication for incoming calls. This addition turns the cellphone into a multisensory device that now is not only heard or seen but also felt, even if the feeling is only one type of vibration. What unites all the permutations of this historical variation is the embodiment: the cellphone is not looked at, because it is either held near the ear or kept in the pocket, on a holster or in a bag.

The borders between the historical variations are not clear-cut, and so one can find early signs of writing in the handsets of the first historical variation.

For example, StarTAC's screen—like many screens of that time—was LED-based, so it was optimized to display numbers. Yet, some forms of texts were displayed, mainly system notices such as "no message" presented in a compressed way as "NoMssgs," where the digit *5* stands for the letter *s*, for example. With such a presentation of letters, it became evident why it was difficult to show the caller's name on that LED screen, and why the Phone Book stored only phone numbers, without the associated names. The Phone Book contained up to twenty numbers, and the user had to recognize the number and make the hermeneutic work of associating each of them to a name. These are the early stages of personalization of the phone, the process in which it has become a personal assistant that relieves us from the need to memorize.

Had we considered the cellphone from a narrowly defined voice communication perspective, then we would have concentrated on its capability to initiate and receive calls everywhere (provided there is cellular coverage, of course). Indeed, such a perspective explains many changes in the world (performed along with the changes in "I" and in "technology" according to the formula "I-technology-world") like the disappearance of public phones and the emergence of "quite cars" in trains. However, the first historical variation is more than a phone with mobility that enables us to talk while away from our home or office. More profoundly, the cellphone has transformed our reference to homes and offices; the distinction between dwelling and workplace has been melting down. When friends are calling during work hours and colleagues during leisure time, it is no longer clear when I'm "at work" and when "with friends/family."

The second historical variation is focused on short alphabetic texts, such as text messages, calendar entries, notes or names in a phone book. It correlates with "2G," but the focus on the experiences related to it led me to study the practices of writing short texts instead of the new communications it enables. The handsets of this historical variation were smaller than the brick, arriving in a "capsule" configuration, also known as "candybar."[10] Yet, clamshell configurations continued to exist and develop, and hybrids emerged in the form of a sophisticated capsule with a sliding part that reveals the keypad. All these forms were optimized to be held in the hand, and significant efforts were invested in the design of the handsets. In all cases, the screen grew from a single line of LED display to a few lines capable of displaying alphanumeric characters. Early applications were comprised of an alphabetic-oriented calendar, address book and notes, as well as a calculator and simple games like "Snake."

The interaction of these applications with the user entails a different embodiment from that of voice communications. The cellphone is no longer a ready-to-hand that is out of sight, in the pocket or near the ear. When typing, the cellphone "occupies" the hands; it becomes literally "present-at-hand" (cf. Plant 2003, 53). We can determine from a distance whether a user is skilled with texting by noticing the way he or she holds the cellphone and carries the body toward the device. There is also a different gaze for new users: they need to look at the keypad and screen while typing to make sure the right letter was selected; skilled users look only on the screen to proof-read text as it is being written, much like computer screen writing practices. The central role of the thumb in skilled users' experience led culture researcher Sadie Plant to announce "the thumb generation" (Plant 2003, 53).[11]

Most scholars refer to the cellphone's texting capabilities as a communication method. A different approach is proposed by Maurizio Ferraris who constructs the cellphone as "a writing instrument" (Ferraris 2005). The ability to write short notes, memos, to-do's, calendar entries, or names in a phone book, changes the cellphone's role from merely "Connecting People," as Nokia's slogan declared, to a writing tool. These short texts have nothing to do with communications with other people. They are for the benefit of the user. For Ferraris, the keypad is a key feature of a cellphone that records. The cellphone that records becomes a "memory prosthesis" (cf. Stiegler 1998). It is an extension of our memory, which can contain dozens and hundreds of contacts, unlimited notes and our past and future schedules. Ferraris' view may explain the rise of cellphones with full keyboards (e.g., RIM's Blackberry or Nokia's Communicator), as well as the increase in memory capacities. It can also explain the disappearance of small notebooks and pens from our everyday carry (cf. Farman 2012). However, such a view hardly explains the continuous growth of the screen, which intensifies in the third historical variation.

The third historical variation focuses on multimedia applications that handle maps, pictures, video clips and music, in addition to traditional data types of numbers and texts. The roots of this historical variation can be traced back to the introduction of Nokia's Communicator in 1996, the first cellphone to integrate voice, text communication, emails, Internet access and 70 applications. Many smart phones followed (like RIM's Blackberry), providing a "corridor" to the Internet and the plethora of information available there. The "poster child" of this historical variation is Apple's iPhone, which was introduced in January 2007, eleven years after Motorola's announcement of the StarTAC. At the time of the launch, iPhone contained a few dozens of applications, but a year later the "App Store" was launched, opening a flood of applications for the cellphone. Obviously, multimedia applications require a relatively large screen, larger than few lines of text. Hence, Apple was proud of the iPhone's screen size (which, few years later, was further extended as iPad). But it was not a screen like other cellphones' screens—it was a touch screen. A touch screen means that it is not only viewed, but also touched (though the feeling is—like StarTAC's vibration—single dimensional, as it always produces the same sleek smooth feeling, regardless of the presented content). This dictated a new interface to the user. Operation was not by pushing buttons but rather by touching the screen usually with the pointing finger. While the second historical variation was based on the thumb ("the fifth finger" according to Gerard Goggin (2004)), users of a touch screen employ the pointing finger (with the addition of the thumb in cases where two fingers are needed, like for zooming in and out). The finger not only presses buttons, but also browses over lists, selects and zooms in and out. That is why iPhone came with only one button, a big difference when compared to cellphones with a keypad, or smartphones with miniaturized keyboard (e.g., Blackberry handsets).[12]

One of the innovations brought by the iPhone was the embedded sensors, like those participating in the adjustment of the display to the holding position (near the ear or being looked at) and the lighting conditions (dark or bright). These sensors form the basis for the fourth historical variation which is about sensing the environment and deciphering it. It is the latest historical variation and hence is indebted to the "histories of the future," though more and more features can be experienced these days.

The cellphone of this last historical variation is much closer to the body of the user. It is embodied as glasses, a bracelet, or a ring, to name some of the artifacts launched in 2012–2014. Many of these transmit data to the cellphone that is nearby, some operate independently. It is as if the cellphone is dismantled and disintegrated, a becoming-rhizome that evolves in various directions. By so evolving, the cellphone interacts with the user in many micro-relations, and the interface is no longer limited to making conversations, entering texts or looking at the screen. Now the cellphone participates in measuring the heartbeat, recording movements or seeing the world from the user's perspective.

These developments occur in parallel to developments in cellular standards known as "4G" or "5G" but they involve other communication standards that handle short-range communications such as Bluetooth. The main aim is not the communication with other people (which was the only aim in the first variation and an important one in the second), but the "self-deciphering" of the user, including the world that surrounds her. That is why the cellphone of this variation has an integrated camera, embedded GPS, and internal sensors for light, movement and the like. All these additions participate in the deciphering of the environment. When compared to the third variation, it appears that such a deciphering should preferably be automatic and so even the single button of iPhone becomes redundant. Indeed, setting requires a full keyboard—real or virtual—but the ongoing operation is seamless as it runs in the background.

Taking the future as the departure point for the study of the cellphone "colors" the historical variation analysis. At this point the StarTAC that was chosen to exemplify the first variation becomes the first wearable cellphone, as its designers heralded it. The second variation contributes the important aspects of visibility and "presence-at-hand." And the third variation enriches with the multimedia elements that mediate the world not only as voice and text but also as pictures, videos, maps and many other forms and formats. It provides a richer experience in terms of both embodiment and hermeneutics.

VARIATIONS AND GENEALOGY

If one examines communication patterns of users in isolation from the mediating communication technologies, the result might be discontinuities, showing—for example—that text messaging emerged out-of-nowhere. But if one examines the functioning of both technologies and users at the same stroke, as performed in the historical variation analysis above, the result is continuous. For example, the analysis clearly shows that certain textual strings appeared on the early LED screens, even though these screens were optimized to the presentations of numbers. This analysis positions messages as a technological potential that existed before communicating-texting practices were developed. The continuities blur the borders between the variations and so applications can be spotted before the emergence of iPhone or other smartphones and pictures were taken as early as the second historical variation. The postphenomenological methodology of historical variations assists in revealing the continuities. Yet, it is not unique. Historians of technology like George Bassala (1988) have advocated for continuity in the evolution of various technologies. The transistor, for example, did not occur in a vacuum. There is—Bassala claims—a continuousness in the development of technologies, and "disruptive" concepts such as the "genius inventor" should be rejected.

A parallel understanding can be found in Colin Koopman's (2013) account of Foucault's intellectual evolution from archaeologies to genealogies. Archaeologies, as an earlier Foucaultian methodology, serve to investigate the past as layers, and the movement between these layers is conceived in terms of ruptures and discontinuities. The discontinuities should be attributed, according to Koopman, to the mode of investigation—archaeological analysis examines a one-vector phenomenon. Genealogies overcome the discontinuity by examining two vectors; they deal with "temporal multiplicity" and "multiple temporalities" (p. 31). This form of genealogy is an extended archaeology, a supplement to it. Put differently, archaeology models isolated *formations* (epistemes in Foucault's terminology) separated by discontinuities and ruptures, while genealogy describes historical processes in terms of trans*formations* that unite into continuities. The objects of the genealogical analysis change endlessly (When Foucault makes this assertion it is radical, but in the context of technology, especially digital technologies, it is trivial). Koopman further maintains that the two vectors in Foucault's genealogies are power and knowledge and that Foucault examines the complexities of these two vectors and analyzes their relations.

If archaeology evolves along a single vector and genealogy deals with two, is there a possibility to study three vectors? Koopman leaves this option as theoretical. I would like to suggest here that postphenomenology studies three vectors, but due to the complexity of the interrelations, the vectors are more straightforward. I suggest that instead of examining the relations between power and knowledge, postphenomenologists investigate the relations of I-technology-world. Like power and knowledge, which are "separate element[s] not identical . . . and yet in constant interaction" (p. 35), there are mutual relations of co-constitution[13] between I, technology and world. Vectors are not studied in and of themselves: Foucault does not attempt to produce a general theory of Power; Postphenomenology does not attempt to produce a general theory of Technology. They both explore relations.

In Koopman's construction of Foucault's genealogies, there is a tight link between genealogy and histories of the present. He writes, "if we are to reconstruct our present so that it may yield better futures, we first need a grip on the materials out of which our present has been constructed in the past" (p. 12). I like this explanation because it ties between the present and the past, and at the same time ties both to the future. In the context of contemporary technologies, the future, I argue, has a more important role. Because the present moves faster, we need to look ahead into the future. When we look ahead, we reconstruct our futures so they yield better presents. At the same time we reconstruct the past in order to identify the potentialities for these presents and futures. Koopman explains, "archaeology asks about what has existed in the past, and without concern for how that which existed came into being. Genealogy asks how that which existed emerged into being in the first place" (p. 40). I suggest shifting the focus of the genealogical analysis from the past to the present and paraphrase Koopman so that the analysis asks how that which is existing is becoming and developing.

The postphenomenological methodology of variations and Foucaultian genealogy share some major characteristics and the similarities are greater than the differences. First is the movement along an articulation. They both refer to a certain phenomenon and review it through the study of its articulated components, may they be historical periods or certain practices in a given time frame. They additionally share the tendency to perform this articulation on the time vector thereby analyzing the history of that phenomenon and how it came into being. Second, variations, like genealogy, do not ask what a specific technology is, but perform an investigative inquiry into how that technology came into a certain mode of being, that is, how it works with humans and how humans work with that specific technology. Third is the shared target. Koopman suggests, after Paul Rabinow, that genealogy can be aimed at "mak[ing] visible what is emerging" (p. 39). The equivalent in postphenomenology can be the invariants, frequently concepts that illuminate certain aspects that appear in all the variations.

EPILOGUE

Many of the press releases announcing new cellphones use words like "revolution" to convey the message that there has been nothing similar to the new handset before the launch. In philosophical terms, they imply Hegelian *Aufhebung*, which denotes "a relationship between two terms where the second at once annuls the first and lifts it up into a higher sphere of existence" (Spivak 1997, xi). If we look at the technological evolution through the lenses of surprising revolution, there is hardly a way to predict the future. But the analysis of the historical variation shows how each variation has its roots in previous variations, thereby providing a factual support to Bassala's continuity and Foucault's genealogies.

Neither is it accurate to describe the cellphone's evolution in terms of problem-solution as media scholar Paul Levinson does. He names the cellphone "a remedial medium" because, he claims, it remedies the problem of being bound to the wires at home (Levinson 2004, 9). Therefore he concentrates on mobility in the sense of out-of-home-or-office, which is equivalent in the analysis conducted here to the first historical variation. This explanation is not necessarily suitable for other developments such as texting, the built-in camera or the embedded GPS. The addition of texting, camera or GPS can barely fit into a problem-solving scheme. Moreover, the reversal of means-end as pointed by Heidegger (1977, 12) and Hans Jonas (2003, 196) positions the description of "remedial technologies" as anachronistic and irrelevant for postmodern technologies, such as the cellphone.

Reviewing the cellphone's evolution, it appears that many of the major characteristics of a historical variation do not replace those of its predecessor, but rather add to it. The "aggregate" allows the coexistence of functionalities and practices. This way, functionalities are added not because of a "need," but just because they can be added (e.g., camera). The evolutionary path of the cellphone does not entail the elimination of previous functionalities and practices. Texting did not eliminate voice communications, and voice is still relevant even for the iPhone.[14]

Based on the description of the evolution of the cellphone, I suggest it is more accurate to refer to a sedimented model that works in layers, where each new layer adds to the previous, and in a delicate process shifts the focus to a new type of activity (cf. Dodd 2004, 67–70). This scheme may also explain why it is difficult to determine the starting and ending point for each historical variation. The case of the cellphone demonstrates how our being-in-the-world has been changing from "connecting people" through a writing machine serving as a memory prosthesis to a true multiplicity device. Based on the sedimented scheme, we can detect a possible direction of a technology that senses our environment, and augments it by revealing aspects which could not have been available to us otherwise: price comparison of a product in the supermarket, location of an ATM or gas station behind the corner (or two blocks away).

Technologies are difficult to predict. They are a kind of unintended consequences, and so are their impacts on their users and the world. Using postphenomenology, the near-term consequences become clearer, because this kind of philosophy of technology keeps opening new horizons for future developments. While philosophy of technology is often considered as not only theoretical but also intricate, this chapter may reveal the useful sides of such philosophy. This is the meaning of "how does one do phenomenology, rather than discourse about it or engage in internecine interpretations" (Ihde 2003).

NOTES

1. For Blackberry's wrong update in Israel in October 2010 see http://192.118.73.5/hasite/spages/1191861.html (in Hebrew; accessed 8 April 2012). This bug was not limited to Blackberry—see Graham Harman's blog post of 29 April 2011on iPad's failure to update time in Egypt in 2011 (http://doctorzamalek2.wordpress.com/2011/04/29/an-hour-earlier-than-i-thought/, accessed 8 April 2012); or PCWorld of March 2011 on iPhone's wrong update of daylight saving in the US (http://www.pcmag.com/article2/0,2817,2381901,00.asp , accessed 8 April 2012).

2. In spite of their major role, and although they are close to us, we do not think of our everyday tools, as long as they operate according to preconceived functions. They "withdraw" to the background, and become ready-to-hand. Not noticing everyday tools is termed by Heidegger a ready-to-hand (*zuhanden*) relationship. Once a tool, such as the hammer, is broken, unusable or absent, he maintains, we become aware of its presence or absence, and it becomes present-at-hand (*vorhanden*). The ready-to-hand and the present-at-hand are the basic relations we maintain with the world, according to Heidegger (1996). In contemporary context, the cellphone is an up-to-date illustration of the ready-to-hand relations we have with technologies, and through them—with the world. The hammer and the cellphone are parts of the background of everydayness that is taken for granted.

3. There are also the cultural variations, where functionally identical artifacts become different technologies when they are embedded in different cultural structures (Ihde 1990; Selinger 2009). Cultural variations reveal the multiple ways in which technologies are embraced by various people in different geographies. The combination of the historical with the cultural may result in a too complex model to fit in an article.

4. It is an aggregate, to use Deleuzian terminology, that is—a behavior which is the sum of the individual behaviors of the various parts. The aggregate is opposed to the assemblage which is a behavior that is not deducible from its parts (DeLanda 2006).

5. It is covered by the standards Global System for Mobile (GSM) and Code Division Multiple Access (CDMA). GSM was originally European, and CDMA was developed in the US, and adopted by ITU. These standards were deployed around the world, so that one country could have two cellular networks operating one in GSM and the other in CDMA. These 2G networks were dominant in the 1990s, and are still operating in various versions into the twenty-first century (see Goggin 2006, 32; for a detailed review on the emergence of GSM see Weber 2009, 121–123.

6. For example, Plant 2003; Castells, et al. 2004; Ling 2004; Horst and Miller 2005; Lemish and Cohen 2005; Ling 2008; Ling and Campbell 2009; Selinger 2009; Schroeder 2010; Rosenberger 2012.

7. For example, Wolf 2000; Ling 2004, 145–168; May and Hearn 2005; Igarashi, Takai and Yoshida 2005; Drouin and Davis 2009; Weber 2009.

8. For example, Levinson 2004; Goggin 2006. Cf. MacKenzie 2010.

9. Yet, it can be easily extended to earlier forms like citizen band radios which mobile telephones that were installed in cars in the 1970s mostly in Germany (see Weber 2011).

10. See for example the gadget website shopper.com, which classifies cellphones by style into candybar, flip, dual screen, slider and swivel. For a comparison between candybar and clamshell see Richard Hantula, Debra Voege, How Do Cell Phones Work? (New York: Infobase Publishing, 2010), p. 21.

11. Another aspect of the thumb generation is related to a physical change in our body. Our hands are changing as demonstrated by the syndrome of BlackBerry thumb, a form of repetitive strain injury caused by the frequent use of the thumb(s) to press buttons on the miniaturized keyboard of smart phones.

12. Yet, the name "touch screen" should not be mixed with the sense of touch. A haptic technology would enable a user to feel and interact in a different way from that offered by a keyboard, no matter if the keyboard is virtual, miniaturized or "regular." The fact that a touch screen uses a tactile organ (i.e., the fingers) does not turn it into a truly haptic technology. The screen remains smooth and sleek, regardless of the application and the content (cf. Paterson 2007).

13. The postphenomenological term of co-constitution is what Koopman calls "coproduction" (37).

14. It is in line with Bassala's concept continuity in the history of technology, which is contested against the discourse of the genius innovator and discontinuity. But Bassala assumes single mode continuity, and does not develop how such continuity works.

REFERENCES

Bassala, George. *The Evolution of Technology*. New York: Cambridge University Press, 1988.

Castells, Manuel, Mireia Fernandez-Ardevol, Jack Linchuan Qiu, and Araba Sey. *The Mobile Communication Society: A cross-cultural analysis of available evidence on the social uses of wirelss communciation technology.* A research report prepared for the International Workshop on Wireless Communication Policies and Prospects, Annenberg School for Communication, University of Southern California, Los Angeles: Annenberg Research Network on International Communication, 2004.

Ceruzzi, Paul. "Moore's Law and Technological Determinism: Reflections on the History of Technology." *Technology and Culture* 46, no. 3 (2005): 584–593.

DeLanda, Manuel. *A new philosophy of society: assemblage theory and social complexity.* London, New York: Continuum, 2006.

Dodd, James. *Crisis and reflection: an essay on Husserl's Crisis of the European sciences.* Dordrecht: Kluwer Academic Publishing, 2004.

Dorrestijn, Steven. "Technical Mediation and Subjectivation: Tracing and Extending Foucault's Philosophy of Technology." *Philosophy and Technology* 25 (2012): 221–241.

Dreyfus, Hubert L., and Paul Rabinow. *Michel Foucault: Beyond Structuralism and Hermeneutics.* Chicago: University of Chicago Press, 1982.

Drouin, Michelle, and Claire Davis. "R u txting? Is the Use of Text Speak Hurting Your Literacy?" *Journal of Literacy Research* 41, no. 1 (2009): 46–67.

Farman, Jason. *Mobile Interface Theory: Embodied Space and Locative Media.* New York and London: Routledge, 2012.

Ferraris, Mauricius. "Where are you? Mobile ontology." *Seeing, Understanding, Learning in the Mobile Age.* Budapest, 2005.

Foucault, Michel. *Discipline & Punish: The Birth of the Prison.* Translated by Alan Sheridan. New York: Random House, 1995.

Goggin, Gerard. *Cell Phone Culture: Mobile Technology in Everyday Life.* London and New York: Routledge, 2006.

———. "Mobile Text." *Media Culture* 7, no. 1 (Jan 2004): http://journal.media-culture.org.au/0401/03-goggin.php.

Heidegger, Martin. *Being and Time.* Translated by Juan Stambaugh. Albany: State University of New York Press, 1996.

———. "The Question Concerning Technology." In *The Question Concerning Technology, and other essays*, translated by W. Lovitt. New York: Garland Pub., 1977.

Horst, Heather, and Daniel Miller. "From Kinship to Link-up: Cell Phones and Social Networking in Jamaica." *Current Anthropology* 46, no. 5 (December 2005): 755–778.

Igarashi, Tasuku, Jiro Takai, and Toshikazu Yoshida. "network Development via Mobile Phone Text Messages." *Journal of Social and Personal Relationships* 22, no. 4 (2005).

Ihde, Don. *Technology and the Lifeworld: from Garden to Earth.* Bloomington and Indianapolis: Indiana University Press, 1990.

———. *Expanding Hermeneutics: Visualism in Science.* Evanston: Northwestern University Press, 1998.

———. *Postphenomenology - Again?* Working Papers from Centre for STS Studies, University of Aarhus, The Centre for STS Studies, 2003.

———. *Postphenomenology and Technoscience: The Peking University Lectures.* New York: State University of New York Press, 2009.

———. *Heidegger's Technologies: Postphenomenological Perspectives.* New York: Fordham University Press, 2010.

Jonas, Hans. "Towards a Philosophy of Technology." In *Philosophy of Technology: The Technological Condition*, by Robert C. Scharff and Val Dusek, 191–205. Malden, Oxford, Victoria: Blackwell Publishing, 2003.

Koopman, Colin. *Genealogy As Critique: Foucault and the Problems of Modernity.* Bloomington: Indiana University Press, 2013.

Lemish, Dafna, and Akiba A. Cohen. "On the Gendered Nature of Mobile Phone Culture in Israel." *Sex Roles* 52, no. 7/8 (April 2005).

Levinson, Paul. *Cellphone: The Story of the World's Most Mobile Medium and How It Has Transformed Everything!* New York, Basingstoke: Palgrave Macmillan, 2004.

Ling, Rich. *New Tech, New Ties: How Mobile Communication Is Reshaping Social Cohesion.* Cambridge and London: The MIT Press, 2008.

———. *The Mobile Connection: The Cell Phone's Impact on Society.* Morgan Kaufman, 2004.

Ling, Rich, and Scott W. Campbell. *The Reconstruction of Space and Time: Mobile Communication Practices.* New Brunswick: Transaction Publishers, 2009.

MacKenzie, Adrian. *Wirelessness: Radical Empricism in Network Cultures.* Cambridge: MIT Press, 2010.

Manovich, Lev. "The Poetics of Augmented Spacce." *Visual Communications* (SAGE) 5, no. 2 (2006): 219–240.

Marx, Karl, and Friedrich Engel. *The German Ideology.* International Publishers Co., 1970.

May, Harvey, and Greg Hearn. "The mobile phone as media." *International Journal of Cultural Studies* 8, no. 2 (2005).

Paterson, Mark. *The Senses of Touch: Haptics, Affects and Technologies.* Oxford and New York: Berg Publishers, 2007.

Plant, Sadie. *On the mobile: the effects of mobile telephones on social and individual life.* Motorola, 2003, http://www.motorola.com/mot/documents/0,1028,333,00.pdf.

Postman, Neil. *Technolopoly: The Surrender of Culture to Technology.* New York: Randon House, 1992.

Reading, Anna. "The Mobile Family Gallery? Gender, Memory and the Cameraphone." *Trames* 12, no. 62/57 (2008): 355–365.

Richardson, Ingrid. "Mobile Technosoma: Some phenomenological reflections on itinerant media devices." *Fibreculture* , 2005: http://six.fibreculturejournal.org/fcj-032-mobile-technosoma-some-phenomenological-reflections-on-itinerant-media-devices/.

Richardson, Ingrid, and Wilken Rowan. "Haptic Vision, Footwork, Place-making: A Peripatetic Phenomenology of the Mobile Phone Pedestrian." *Second Nature* 2 (November 2009): 22–41.

Rosenberger, Robert. "Embodied technology and the dangers of using the phone while driving." *Phenomenology and the Cognitive Sciences* 11, no. 1 (2012): 79–94.

Schroeder, Ralph. "Mobile phones and the inexorable advance of multimodal connectedness." *New Media & Society* 12, no. 1 (2010): 75–90.

Selinger, Evan. "Towards a reflexive framework for development: technology transfer after the empirical turn." *Synthese*, 2009: 377–403.

Spivak, Gayatri Chakravorty. "Translator's preface to Of Grammatology, by Jacques Derrida." In *Of Grammatology*, by Jacques Derrida, ix–lxxxix. Baltimore: John Hopkins University Press, 1997.

Stiegler, Bernard. *Technics and Time, 1: The fault of Epimetheus.* Translated by Richard Beardsworth and George Collins. Stanford: Stanford University Press, 1998.

Weber, Heike. "Consumers as Innovative Actors? The Role of Users in the Shaping of German GSM Telephony." *Le Movement Social* 228, no. 3 (2009): 117–130.

———. "Mobile Electronic Media: Mobility History at the Intersection of Transport and Media History." *Transfers* 1, no. 1 (2011): 27–51.

Wolf, Alecia. "Emotional Expression Online: Gender Differences in Emoticon Use." CyberPsychology and Behavior 3, no. 5 (2000): 827–833.

Chapter 4

On Postphenomenology and the Postcolonial

Srikanth Mallavarapu

Don Ihde brings together elements of hermeneutics, phenomenology, and pragmatism in his attempt to create a philosophical approach that can engage in a substantive way with the technoscientific world that we inhabit. This chapter brings Ihde's work into a more explicit conversation with postcolonial theory, a conversation that I believe shows the strengths of Ihde's approach while also raising questions about power and representation in his postphenomenological framework. I argue that Ihde's rejection of technological and social determinism as well as utopian and dystopian approaches to technology is particularly important, given that these keep reappearing in debates about science, technology, and modernity in the postcolonial context. Ihde's refusal to see science and technology as a purely European/Western enterprise also allows for the possibility of engaging with the trajectory of specific technologies in different parts of the world and exploring alternate or hybrid modernities through a methodology that focuses on embodied experience and material practices without ignoring the importance of cultural context. The final section of this chapter will focus on the politics of representation in Ihde's framework, suggesting that it is possible to draw on insights from postcolonial theory in order to refocus the hermeneutic lens on the way that postphenomenological narratives are themselves created in order to allow for a deeper political engagement and greater situatedness.

Don Ihde's extensive body of work emphasizes the importance of embodiment, materiality, instruments, and practices. The focus on practice that comes from the pragmatist tradition is linked up with phenomenological variational theory, with careful attention given to embodied experience and human-technology interactions in the lifeworld. As one of the philosophers associated with the "Empirical Turn" in the philosophy of technology (Achterhuis 2001), Ihde's work is a "step away from generalizations about *technology uberhaupt* and a step into the examination of *technologies in their particularities*" (Ihde 2009: 22). This focus on specific technologies in specific contexts offers a challenge to traditional approaches to technology, which involve either social or technological determinism. Theorists who are social determinists argue that social factors (more specifically, the elite as represented by a technocratic bureaucracy) determine in crucial ways the way technology is implemented in order to maintain the established hierarchy. The nature of the technology being implemented is incidental to the analysis—the assumption is that technologies are neutral, and are just put to use to maintain existing power structures.

In contrast to the social determinists, the technological determinists argue that it is technology that determines social developments. Ihde gives the example of James Burke and Lynn White's work on the development of the stirrup. This analysis in a systematic way traces the implications that the stirrup had on the development of later medieval technology such as lances, saddles, and armor (Ihde 1990: 5–6). The assumption that these analysts make is that there is a certain inevitability to the development of technology, which makes it stand outside the specific historical and cultural context in which it is developed. Ihde rejects both social and technological determinism, suggesting instead that we pay close and careful attention to the particularities of technologies, acknowledging that these technologies are embedded in cultures.

Ihde also points out that approaches to technology tend to be either utopian or dystopic. The utopian approach to technology was very common in the nineteenth century, and traces of it can still be found in scholars like Meera Nanda who see science and technology offering the best chance of addressing the multitude of problems the world faces (Nanda 2003). Given the history of technology and industrialization in the twentieth century, the century of Hiroshima and Nagasaki and Bhopal, it seems hard to hold on to this utopian perspective, but as Ihde points out, the advocates of Artificial Intelligence are technological utopists, and there are others as well—for example, those who believe that developments in the mapping of the genome will hold the key to almost every medical problem or those who believe that large-scale geo-engineering is the best way to tackle climate change.

In contrast, many philosophical approaches to technology from the twentieth century to the present day have tended to be dystopian. Heidegger's work on technology has been extremely influential in this regard (Heidegger 1977). Heidegger made the important move of suggesting that technology precedes science ontologically, instead of the standard assumption that technology follows from developments in the scientific domain. As Ihde puts it, "For Heidegger, technology's ontological priority over science shows that all science must be, in my term, technoscience" (Ihde 2009: 35). While Ihde does agree with Heidegger's reversal, he distances himself from Heidegger's negative characterization of modernity as being defined by a technological approach that frames the world as a mere standing reserve (Bestand). In the 1960s Herbert Marcuse and Jacques Ellul argued that "technology was equatable with certain calculative ('rational') techniques which in turn become not only dominant but total, thus making technology as culture absolute" (Ihde 1990:6). With the gradual growth of the environmental movement around this time, there were critiques of technology rejecting its "artificiality" in favor of a pristine nature, which was romanticized. Ihde rejects both the utopian and dystopian approaches to technology in his attempt to build a more nuanced theoretical framework that also resists determinism. It is through an understanding of technology and its contexts that one gains an understanding of nature and our technologically mediated lifeworlds.

In Ihde's work, we see a modified version of the Husserlian notion of the lifeworld (Husserl 1970). Husserl's reading of Galileo offers an interesting distinction between the everyday lifeworld which all of us have access to, and the idealized scientific lifeworld, which substitutes the everyday lifeworld for Galileo. Ihde points out that the problem with the Husserlian analysis is that it empties the scientific lifeworld of embodied perception and praxis that constitutes the everyday lifeworld. Merleau-Ponty, in contrast, offers a sophisticated analysis of the correlation between "lived" body and lifeworld perception in phenomenological terms (Merleau-Ponty 1962). Ihde argues that "it is out of this sensitivity toward body-world correlation that Merleau-Ponty anticipates a different role for technologies within the realm of perception and praxis. The lived or virtual body as an experienced bodily spatiality can be 'extendible' through artifacts" (Ihde 1990: 39). The famous examples of such artifacts are the feather in a hat, and the blind man's cane. Following this phenomenological tradition of studying embodiment and its extensions, Ihde argues that to understand technology, we need to understand the specific context of the technology and this can be done only by paying attention to the bodily praxis involved. Ihde then goes on to broaden his project by addressing the question of how "cultures embed technologies" (Ihde 1990: 124). The study of embodied praxis and the use of technologies in specific historical and cultural contexts is central to Ihde's project and this postphenomenological approach can help avoid both determinism as well as utopian/dystopian tendencies in postcolonial debates about the negotiation of modernity.

Science and Technology Studies (STS) and postcolonial theory are both interdisciplinary configurations which have attempted to interrogate terms like modernity and rationality at a fundamental level. Postcolonial theory explicitly has to deal with the encounter with modernity in the context of the colonial encounter. Work in Science and Technology Studies has important implications for the study of difference because the demarcation between scientific and nonscientific domains has served as a significant marker of difference, and indeed as a justification for colonialism, the argument being that the colonial encounter would bring the benefits of the Enlightenment and modernity to the "underdeveloped" parts of the globe (Adas 1989). While postcolonial theory explicitly has to deal with the problems of difference in the context of the complex negotiation of modernity, work in Science Studies has important implications in this regard because the demarcation between scientific and nonscientific domains serves as a significant marker of difference. Paul Feyerabend argues in his provocative way that the myth of scientific method serves to privilege western modes of knowing over non-western modes (Feyerabend 1975). Bruno Latour makes the claim that the idea of modernity is propped up by an artificial demarcation between the realm of nature (studied by science) and the realm of culture, and that this then results in the demarcation between the West and non-West (Latour 1993). There is no denying that the attempts to demarcate the border between scientific and nonscientific domains have resulted in the delegitimization and in some cases the silencing of nonmodern ways of being in the world.

The word "science" brings with it the dream of objective knowledge, a celebration of critical thinking and rationality that liberates humankind from dogmas and superstitions. Many feminist and postcolonial scholars and activists have been inspired by these ideals, arguing that women and minority groups who had in the past been excluded from the universalizing gaze of science needed to be brought into the fold. This valorization of science and technology is closely linked to the rhetoric of progress and development—in so-called "developing" countries, science and technology are seen as crucial elements in the fight against poverty, disease, and hunger. The scientific worldview is also seen as providing a crucial resource in the struggle against oppressive local traditions.

In contrast to this approach, other scholars have argued that the myth of an objective science only serves to mask the fact that the history of modern science is inextricably linked to patriarchal, capitalist, and (neo)colonialist structures. In the Indian context, the work of intellectuals and activists like Mahasweta Devi has exposed the violence that has been committed in the name of progress and development (Devi 1997). It is usually the subalterns (the people on the fringes of modernity) who bear the brunt of this violence. It is also impossible to ignore the problematic links between modern science and technology and multinational corporations—the Bhopal gas tragedy which occurred in 1984 when the lethal methyl isocyanate gas leaked out from a Union Carbide factory is a tragic reminder of this. Making arguments that parallel those made by Marcuse and Ellul, Ashis Nandy has argued that the simple valorization of science and technology does not take into account the complex ways in which the technocratic worldview, assuming the authority of "science," has been itself caught up in deeply oppressive structures

(Ellul 1964, Marcuse 1964, Nandy 1990). In this particular debate about the nature of "development" in the postcolonial world, both the votaries of technology and its opponents see modern science and technology (which is seen to be Western) as clashing with local traditions. However, the complicated contemporary global terrain with flows of capital, products, technologies, and labor with the ongoing debates about markets, intellectual property rights, patenting and protectionism etc. clearly demonstrates that any simplistic model that divides the West and non-West is inadequate.

It is not only in the contemporary world that we inhabit that the division between the modern West and the nonmodern non-West becomes problematic. In his critique of Latour's account of the divide, Gyan Prakash points out in *Another Reason* that the

> identity and authority [of Western modernity] were forged on the stage of colonial and imperial domination . . . is it possible to deny the *simultaneity* of the formation of Western scientific disciplines and modern imperialism? One thinks for example, of connections between the West's global expansion and the formation of the disciplines of ethnology, political economy, botany, medicine, geology, and meteorology. (Prakash 1999: 12–13)

In addition to historians like Prakash who challenge the idea of Western modernity being formed in Europe and then exported to the far reaches of the globe, there are also historians who argue that we need to pay attention to other non-Western modernities. In a recent review of Pankaj Mishra's *From the Ruins of Empire: The Intellectuals Who Remade Asia*, David Shulman brings up the work of scholars like Velcheru Narayan Rao and Sanjay Subrahmanyam who (with Shulman) have argued that a form of modernity was taking shape in South India by the end of the fifteenth century, prior to the arrival of Vasco De Gama:

> Thus we find, with particular prominence, the concept of an autonomous, subjective individual, responsible for his or her fate; a new theory of romantic love; the development of literary fiction as a privileged literary technique; a vogue for skepticism and realism, seen as informing the pragmatics of everyday life; the emergence of a cash economy and the conceptual revolution that rapid monetarization entails; the appearance of a bold, full-throated, unfettered female voice; and a new concept of nature as a rule-bound domain, separate from the human and amenable to disciplined observation and extrapolation. An innovative economic model of the mind, centered on the imaginative faculty, came to define the meaning of being human. (Shulman 2012)

Scholarship like this suggests that it is not just in the current era of globalization when the idea of a simple divide between the modern West and the nonmodern non-West breaks down.

Ihde's framework rejects the stereotypes of a technologically advanced modern West that is defined against a "tradition-bound" nonmodern non-West. Ihde points out that rudimentary forms of technology have been used by human societies in contexts that have been assumed to be "natural" and technology free. In a move that is similar to the one made by Haraway in her cyborg manifesto (Haraway 1991), Ihde points out that the dream of an innocent and pristine nature untouched by technology is just that—a dream. In addition to refusing to romanticize the preindustrialized world as being more "natural" by insisting on the long history of tool-use, Ihde has also challenged what he has called in a letter to *Science* "the traditional and parochial display of Eurocentrism regarding the history of science and technology," pointing out the Chinese had invented mechanical clocks and the Koreans had developed metal moveable type well before the Europeans (Ihde 2000: 803). In *Postphenomenology and Technoscience*, Ihde points out that with science and technology becoming increasingly identified with Europe, the Chinese origins of printing, gun powder, and the compass were quietly forgotten or dismissed as not the products of a truly scientific method (Ihde 2009: 50). Ihde's interest in the multicultural history of science and technology goes beyond just setting the historical record straight—his approach adds a cultural dimension to the trajectories technologies can take and the phenomenological variations that are possible. However the question that remains is the following: Once these binaries (West/non-West and modern/nonmodern) are broken down or at least problematized, is there still room in postphenomenology to offer a rich account of significantly different ways of being in the world or is embodiment flattened to the standard phenomenological body switching through different cultural contexts? How can a methodology that focuses on human-technology-world interactions work to also incorporate larger economic and political analyses that examine the impact of colonialism or capitalism and multinational corporations or state policies? A couple of examples used by Ihde in *Technology and the Lifeworld* allow us to explore these questions.

Ihde demonstrates the limitations of the standard model of technology transfer by describing the first contact between Australian gold prospectors and New Guineans in 1930. Gifts such as knives and axes were clearly appreciated by the New Guineans, but rifles did not evoke the same response, even when their use was demonstrated. It was only later, when the New Guineans realized that the Australians were not gods, but all too human, that the rifles were seen as something worth possessing. This example shows the dynamics of power between the groups and demonstrates the importance of context and praxis—it was because the New Guineans had tools similar to the axes and knives that they could easily fit the new steel knives and axes into an already existing praxis. It needed a little more work to fit the rifle into a praxis—as a weapon that could kill from a distance. Ihde points out the implications of the "gifts":

> the New Guinean is receiving more than a mere artifact. He is potentially receiving a set of cultural relations—in this case, relations that will create a situation of dependency upon the incoming culture . . . Just as there is no such thing as "an" equipment, neither is there an equipment without its belonging to some set of culturally constituted values and processes. As the steel axe and knife become more desirable and since their mode of production is not available to the New Guinean, the new objects must be attained by some form of exchange, the histories of which are multiple enough not to recite here. (Ihde 1990: 126)

Another brief example that Ihde gives is that of the Pulawateans, South Pacific navigators who used wave patterns to steer before being introduced to the compass. Ihde points out that the Pulawateans used the compass to steer in a straight line, without incorporating the larger theoretical framework of directions or the idea of latitudes and longitudes. The use of the compass resulted in the Pulawateans gradually de-skilling themselves from the more complex wave-reading skills. In the examples being discussed, embodied praxis is given precedence, and it is clear that simple models of technology transfer are inadequate to account for the ways in which technologies are culturally embedded; for example, Ihde is very clear that "the compass has a different 'being' in Pulawatean use than in Western use" (Ihde 1990: 126). Ihde's refusal to see science and technology as a purely European/Western enterprise allows for the possibility of engaging with the trajectory of specific technologies in different parts of the world and perhaps exploring the possibility of alternative modernities, especially looking at embodied experience and material practices in this context. Ihde's framework challenges the rhetoric of progress and development that sees science/technology/modernity as inherently emancipatory as well as the rhetoric of nostalgia, which sees Western modernity as violent and seeks to reclaim a prelapsarian paradise.

So far, this chapter has focused on the strengths of Ihde's postphenomenological approach. The final section of this chapter will focus on the politics of representation in Ihde's framework. Since postphenomenology foregrounds embodied experience, the question of how it is being represented is a significant one. In many of his books, Ihde presents fascinating and thought-provoking examples of the postphenomenological approach by offering careful analyses of his own embodied experience and the relationship with specific technologies and his lifeworld. In addition to this, Ihde gives examples from his travels across the world and also uses historical examples. It is important to point out here that Ihde's exploration of embodiment is not an account of subjective experience—there is a rigorous examination of phenomenological variations. The postphenomenological approach developed by Ihde takes care to situate technological embodiment in a cultural context. The distinction that Ihde makes between what he refers to as body one and body two is significant here. Body one is the standard phenomenological body, and body two refers to the social and cultural component of embodied experience. If Merleau-Ponty is the philosopher immediately associated with body one, Ihde sees Foucault as the philosopher associated with body two. Not surprisingly, feminist and postcolonial scholars have a tendency to focus on body two. However, Ihde points out that "for there to be a marked cultural body, or body two, there must be a body one that is markable" (Ihde 2002: 70). Even though Ihde acknowledges that it is hard to draw a clear line demarcating body one and body two, the emphasis in this quote suggests that an analysis of power that draws upon Foucault or Haraway is going to be sidelined by an analysis of embodiment that draws more from Merleau-Ponty, because body one is seen as primary. As a postphenomenologist, how are these choices to focus on body 1 or body 2 made? Robert Scharff asks "how Ihde can be so sure that given all the 'polymorphic ambiguities' that 'lie hidden' in our admittedly problematic world, everything can be handled well by those with some phenomenological training and pragmatic good sense?" (Scharff 2006: 137). We also need to ask if the very idea of body 1 standing outside culture is a particular western conception that might in some cases end up erasing markers of difference.

The final chapter of *Postphenomenology and Technoscience* is entitled "Do Things Speak?" In this chapter, Ihde gives the example of a body (later dated through carbon 14 dating to 5300 BP) that was discovered in the Alps. Even though there was no linguistic information available, Ihde demonstrates how a narrative about the corpse can be constructed through various techniques ranging from CT scans to DNA analysis and mass spectroscopy. Through this example of what he calls "material hermeneutics," Ihde answers his own question "How can we hear, or give speech to, that which is silent?" (Ihde 2009: 69–72). While this is a remarkably good example of how nonlinguistic scientific data is interpreted and then constructed into a narrative, the same question can be asked in the context of human beings who are marginalized from political discourse. In fact, this is the central question that animates the field of subaltern studies, a branch of postcolonial theory. While both Ihde's postphenomenological approach and subaltern studies pay attention to questions of embodiment as well as material practices, the basic difference is the explicit normative component in Subaltern Studies, which is fundamentally a democratic project that focuses on voices that have been marginalized by mainstream historiography. Dipesh Chakrabarty, Ranajit Guha, and Gayatri Spivak, for example, are concerned not just with the negotiation of modernity in a historical context, but also with the problems that interpretive social science frameworks bring with them in the attempt to describe subaltern domains (Chakrabarty 2002, Guha 2002, Spivak 1999). Their work points to the limitations of disciplinary discourses predicated on modernity by acknowledging the uncanny ways in which the nonmodern makes its presence felt in a world that is never completely modern.

In *Provincializing Europe*, Dipesh Chakrabarty offers an interesting reading of the scholarship on weaving communities in colonial India. He examines Gyan Pandey's work on the Julahas, which documents the changes in the lifeworld of these communities as a result of British colonial policies (Chakrabarty 2000). Chakrabarty's larger project in the book is to point out the limitations of disciplinary perspectives based on modern frameworks and universalizing categories like Marx's "abstract labor." Chakrabarty places himself broadly in the Marxist tradition, appreciating the critiques of not just capitalism but also colonialism that come from a study of labor and capital, while cautioning that these categories are not universal and need to be situated in specific historical and cultural contexts. He points out that a classical Marxist analysis using such categories fails to acknowledge that for these communities, work and worship could not be separated. Chakrabarty uses the work of Richard Eaton who has collected songs that were sung by weavers as work was being performed that illustrate this complex relationship. These songs from the Sufi tradition compare parts of the human body to the spinning wheel and exhort the weavers to repeat the name of God as they weave, suggesting an explicit linkage between the act of breathing, repeating the name of God, and the act of weaving, with the body itself being described as being tuned and threaded through this process (Eaton 1978: 163–164, cited in Chakrabarty 2000: 81). One can imagine a postphenomenological analysis of the embodied process of weaving, with the focus on the way that the body of the weaver interacts with the spinning wheel to produce cloth. But just like a standard Marxist analysis of labor and alienation fails to capture the complex lived relationship between work and worship for the weavers, an analysis that starts with body one (the standard phenomenological body) and the ways in which this body interacts with the spinning wheel and then adds the cultural/religious dimensions associated with body two later will fail to capture the complexity of embodied experience being described in the songs. The description of the body that is being tuned and threaded makes any easy separation of body one and body two very difficult.

Warwick Anderson and Vincanne Adams' chapter on Postcolonial Studies of Technoscience in *The Handbook of Science and Technology Studies* (3rd ed) makes the case that the postcolonial approach "offers a flexible and contingent framework for understanding contact zones of all sorts, for tracking the unequal and messy translations and transactions that take place between different cultures and social positions, including between different laboratories and disciplines even within Western Europe and North America"

(Anderson 2008: 184). What does the postcolonial approach have to offer to a Marxist historian trying to understand the lifeworld of a weaving community or a postphenomenologist studying Pulawatean fishermen? There has to be an acknowledgment that the discursive and analytical categories and frameworks used by modern academic disciplines are privileged over the categories used by the weavers or the fishermen to explain their own lives. There has to be an increased sensitivity to the ways in which power relations are refracted through representation. There also has to be an acknowledgment that these disciplinary categories and frameworks might not able to capture the richness and complexity of subaltern lifeworlds. What I am suggesting here is not an uncritical privileging of the perspectives or narratives of the marginalized, but rather an increased critical engagement with them, acknowledging both the strengths and analytical power of frameworks like Marxism or postphenomenology as well as the moments of rupture or breakdown, the uncanny moments when the richness of the lifeworld cannot be contained by these frameworks.

Evan Selinger suggests that Ihde's work has an "intrinsically normative" component in that any discussion of technology can alter human-technology-world relations. It also has a "substantive normative commitment" that critically analyzes "the theoretical positions that over-determine and totalize our technologically mediated relations" (Selinger 2006: 104). Selinger also highlights quotes in which Ihde talks about "caringly looking after our technologically textured world" and being "empathetic to a neo-enlightenment set of social and political values" (Selinger 2006: 104). Part of Ihde's reluctance to embrace an explicit normative component seems to be linked to his preference for a descriptive approach rather than a prescriptive approach as well as his reluctance to make statements about Technology as opposed to specific technologies. However, if the goal is to develop a truly situated model of postphenomenology, perhaps the normative component could be made a little more explicit without giving in to pronouncements about Technology that slide into social/technological determinism or utopian/dystopian rhetoric. Ihde's framework is an elegant and powerful one that has already found adherents across the continents. The range and quality of work being produced under the postphenomenological umbrella is proof that even without an explicit normative component, this framework allows for projects that are engaged with questions of politics while paying close attention to embodiment, material practices, and human-technology-world interactions and the ways in which cultures embed technologies. Turning the hermeneutic lens on the way that postphenomenological narratives are themselves created would increase the possibilities for deeper political engagement and greater situatedness; it would also allow us to acknowledge the limitations and assumptions of our disciplinary discourses and open up the possibility of engaging in a more substantial way with different ways of being in the world.

REFERENCES

Achterhuis, Hans. ed. (2001). *American Philosophy of Technology: The Empirical Turn*, trans. Robert P. Crease. Bloomington: Indiana University Press.

Adas, Michael. (1989). *Machines as the Measure of Men: Science, Technology, and Ideologies of Western Dominance*. Ithaca: Cornell University Press.

Anderson, Warwick and Vincanne Adams. (2008). "Pramoedya's Chickens: Postcolonial Studies of Technoscience." In Edward J Hackett, Olga Amsterdamska, Michael Lynch and Judy Wajcman eds, *The Handbook of Science and Technology Studies*, 3rd edition. Cambridge, MA: MIT Press, 181–204.

Chakrabarty, Dipesh. (2000). *Provincializing Europe: Postcolonial Thought and Historical Difference*. Princeton: Princeton University Press.

———. (2002). *Habitations of Modernity: Essays in the Wake of Subaltern Studies*. Chicago: The University of Chicago Press.

Devi, Mahasweta. (1997). *Dust on the Road: The Activist Writings of Mahasweta Devi*, trans. Maitreya Ghatak. Calcutta: Seagull.

Eaton, Richard Maxwell. (1978). *Sufis of Bijapur* 1300–1700*: Social Role of Sufis in Medieval India*. Princeton: Princeton University Press.

Ellul, Jacques. (1964). *The Technological Society*. New York: Random House, Vintage.

Feyerabend, Paul. (1975). *Against Method: Outline of an Anarchistic Theory of Knowledge*. London: New Left Books.

Guha, Ranajit. (2002). *History at the Limit of World-History*. New York: Columbia University Press.

Haraway, Donna. (1991). "A Cyborg Manifesto: Science, Technology, and Socialist-Feminism in the Late Twentieth Century." In *Simians, Cyborgs and Women: The Reinvention of Nature*. New York; Routledge, 149–181.

Heidegger, Martin. (1977). *The Question Concerning Technology and Other Essays*, trans. William Lovitt. New York: Harper Perennial.

Husserl, Edmund. (1970). *The Crisis in European Sciences and Transcendental Phenomenology*, trans. David Carr. Evanston: Northwestern University Press.

Ihde, Don. (1990). *Technology and the Lifeworld: From Garden to Earth*. Bloomington: Indiana University Press.

———. (2000). "Timeline Travails" *Science* Vol. 287 (February 4, 2000), 803.

———. (2002). *Bodies in Technology*. Minneapolis: University of Minnesota Press.

———. (2009). *Postphenomenology and Technoscience: The Peking University Lectures*. Albany: SUNY Press.

Latour, Bruno. (1993). *We have never been Modern*, trans. Catherine Porter. Cambridge, Mass.: Harvard University Press.

Marcuse, Herbert. (1964). *One-Dimensional Man*. Boston: Beacon Press.

Merleau-Ponty, Maurice. (1962). *Phenomenology of Perception*, trans. Colin Smith New York: The Humanities Press.

Nanda, Meera. (2003). *Prophets Facing Backwards: Postmodern Critiques of Science and Hindu Nationalism in India*. New Brunswick, NJ: Rutgers University Press.

Nandy, Ashis. (1990). "Introduction: Science as a Reason of State" in *Science, Hegemony and Violence: A Requiem for Modernity.* Delhi: Oxford University Press, 1–23.

Prakash, Gyan. (1999). *Another Reason: Science and the Imagination of Modern India.* Princeton: Princeton University Press.

Scharff. Robert C. (2006). "Ihde's Albatross: Sticking to a "Phenomenology" of Technoscientific Experience." In Selinger, ed. (2006), *Postphenomenology: A Critical Companion to Ihde*. Albany, SUNY Press, 131–44.

Selinger, Evan. (2006). "Normative Phenomenology: Reflections on Ihde's Significant Nudging" In Selinger, ed. (2006), *Postphenomenology: A Critical Companion to Ihde*. Albany, SUNY Press, 89–107.

Shulman, David. (2012). "The Revenge of the East?" *The New York Review of Books* (October 11, 2012), http://www.nybooks.com/articles/archives/2012/oct/11/revenge-east/

Spivak, Gayatri Chakravorty. (1999). *A Critique of Postcolonial Reason: Toward a History of the Vanishing Present.* Cambridge, MA: Harvard University Press.

Chapter 5

What is Multistability? A Theory of the Keystone Concept of Postphenomenological Research

Kyle Powys Whyte

A strand of literature called postphenomenology continues to build steam in some philosophy of technology and science and technology studies circles. Among its signatures is the use of Don Ihde's take on *multistability* as a framework for interpreting the polymorphic relationships between humans and technologies, or human-technology entanglements. Time and again, in conferences and scholarly exchanges, the following question arises: what exactly is multistability as a concept around which to organize a literature called postphenomenology?

I want to know the answer to this question, and will provide one in the form of a theory. I will argue that there are really two conceptions of multistability at play. First, multistability is an empirically testable hypothesis about how several stable patterns of the same object can be perceived from the first person perspective, and resonates well with previous philosophical observations of the same phenomena (Cerbone 2009). I refer to this as imaginative multistability. Ihde has used this sense to discuss the ways in which technologies have mediated his own perceptions (Ihde 1998, 1990, 1976, 2009).

Most research papers in which the authors identify as postphenomenologists, however, are concerned with investigating from a third person perspective the multistability of technologically mediated practices. Building on imaginative multistability, this second sense, which I call practical multistability, is based on the hypothesis that human bodies and technologies are entangled in lifeworlds and includes sets of concepts and criteria that can be used to describe some of these entanglements. The descriptions are used to shed light on the role of these entanglements in framing our aesthetic, moral, and political values and the possible ways of improving the benefits and sustainability of technology designs and built-environments (Rosenberger 2009; Verbeek 2009, 2008; Selinger 2008; Verbeek 2005; Riis 2008; Forss 2011).

I will conclude with a discussion about how my theory of multistability suggests a trajectory of future research in postphenomenology. Owing to the prevalence of research on practical multistability, postphenomenology should envision itself as an emerging social science that can employ certain descriptive criteria and concepts to describe the interactions among humans, nonhumans, and technologies. Part of its value lies not in philosophy, but with client communities of industrial designers, video game creators, and engineers who can use the results of the studies for designing new technologies and critiquing current designs. The success of postphenomenology depends on whether its philosophical practitioners can (1) develop transparent criteria and concepts that can be used and interpreted across multiple collaborations with nonphilosophers and (2) interactional ability with at least one client community.

IMAGINATIVE MULTISTABILITY

I will begin with the briefest of definitions. Multistability refers to the idea that something can be more than one thing or have more than one stability. For now, we can understand *stability* as an umbrella term that can refer to anything perceived as having a constant pattern, from the constancy of images, to practices, to technologies, and so on. Anything that is stable comes across to us as having at least one of the following: a particular look, a particular way of acting, or a particular use. Multistability indicates that the same object can have more than one such stability without altering its composition.

In *Experimental Phenomenology*, Ihde sets out to analyze "so-called visual illusions and multistable visual phenomena" (Ihde 1977, 67). One example is the drawing of the "the Hallway." When asked, "What do you see," one set of responses says "the figure *is* a hallway appearance," and the other says, "The figure *is* a pyramid appearance" (Ihde 1977, 70). Ihde urges us to "[n]ote carefully . . . that each group actually experiences the [perception] as claimed; they can fulfill or verify this assertion experientially, thereby offering a certain evidence for their assertion" (Ihde 1977, 71). This initial certitude is referred to by Ihde as apodicticity, meaning that we can return again and again to the drawing and verify our assertions.

Apodicticity is actually the pattern one perceives. Two different people can have radically different perceptions of patterns, other things being equal. There is also a third person, the person who has both perceptions, realizes it and, as a result, can switch back and forth between them. This person is fully aware of the two possible patterns in perception of exactly the same image. Ihde describes this viewer as a more ascendant viewer insofar as he or she has moved from "literal-mindedness" to "polymorphic mindedness." The ascent is from being able to have only one perception from being able to see the alternatives.

As Ihde writes,

> With the ascent in level, the viewer of alternatives does not lose apodicticity; he is able to return to both appearances and thus fulfill or verify the evidence of both. But the *significance* of that apodicticity changes. Once both aspects are possible, it is clear that neither the hallway nor the pyramid appearance can claim absoluteness or exhaustiveness for the possibilities of the thing. The [perception] is now seen to contain two possibilities, and two possibilities as variations are relatively more *adequate* than one. (Ihde 1977, 72)

One important point here is that the ascent is not from being mistaken, in literal-mindedness, to being more accurate, in polymorphic mindedness. Rather, ascent moves one to continue to be able to have both perceptions in their apodicticities, but also not to privilege any one of them as more accurate. The polymorphic minded perceives both baseline perceptions as being symmetrical as apodictic perceptions. Each of these baseline perceptions of patterns could be called a stability.

If we have this idea down, then, Ihde shows, we can also perceive the image as a headless robot. This is another stability. And we can, of course, look for more. If we were to run through more variations, what we are doing is actively turning our attention to the many possible stabilities that can be apodictically perceived starting from the one image. But every time we come across a stability we have to note some of the *prima facie* factors that make such a stability appear with apodictic resonance. If we view the image as a hallway, we have the point of view (POV) that we are above it looking into it. If we view the image as pyramid, it is as if we have the POV of flying over it. And if we view it as a headless robot, it is as if we are facing it and it is coming at us. Each stability has, built into it, a POV that conditions its appearance. There is also an implicit cultural horizon. Being able to see the headless robot implies that we are aware of what a robot is. Being able to see the pyramids might only be possible from perspectives like mine, for example—I have never been to the pyramids and imagine them as if I were in a helicopter flying over them. Perhaps someone who lives there has a richer conception of the pyramids and would not see one in the image.

What is important to notice is the weak embodiment suggested by this method of analysis. Though I am accounting for POV and culture, I am doing so in a way that still leaves quite a few questions unanswered about who I am as an observer. There is a sense that I am sort of a hovering observer running these variations through with no particular enduring linkage to the context of any lifeworld (specific meanings of a time and place) even though, like with the case of the headless robot, I am not entirely ignoring culture.

Multistability in terms of images—imaginative multistability—is at the heart of many of Ihde's studies (Ihde 1998, 1990, 1976, 2009). What are the values of investigating the multistability of experience from the standpoint of weak embodiment? The first value is a testable hypothesis that images can be perceived differently. By testable, I mean phenomenologically from a first-hand perspective. This is one way of reading Ihde's notion that "variational analysis gives rigor to experiential analysis" (Ihde 2009, 15–16). The second value is an ever-so-slight evaluative hypothesis concerning the idea that being able to perceive multistability is an ascent. In some contexts, it is preferable to be able to recognize multistability and to avoid literal-mindedness. Recently, there is growing interest in the ethical implications of postphenomenology. If there are any, they start here, with Ihde's ideas of ascent and polymorphic mindedness. It is an important question how much morality we can get out of this. Some, like Evan Selinger, believe that polymorphic mindedness is a ground for evaluating ethical controversies over the interpretation of images, but not the entire ground. Polymorphic mindedness can reveal how some parties have hardened perceptions of images that fuel conflict.[1]

Firsthand testability and the awareness of multistability are the major values of philosophizing about our experiences with images and technologies from the standpoint of weak embodiment. But imaginative multistability is not the only kind that Ihde discusses, and I will turn my attention to the more interesting kind in what follows.

PRACTICAL MULTISTABILITY

Ihde transitions from imaginative multistability to the multistability of practices or what I call practical multistability. He writes in his *Peking University Lectures*:

> At this point I want to make a large leap to an example set now related to *technologies*. While the use of visual "illusions" has the advantage of initial clarity and ease to demonstrate multistability as a phenomenological result of variational analyses, these illustrations also have the disadvantage of being all too simple and all too abstract. This is particularly the case with respect to the weak sense of embodiment in the illusions set. My POV . . . is clear but weak in the sense that I am in a mere "observer" position vis-à-vis the examples. So my next example set will draw from a very ancient, very simple, and very multicultural set of technologies: *archery* (bows and arrows). (Ihde 2009, 16)

When we think of practices, like archery, we have to work with a stronger notion of embodiment. What does this mean to say that considering the multistability of practices necessarily avoids weak embodiment?

Ihde claims that when we think of practices like archery we should "look for variations, embodiment, . . . [and] lifeworld dimensions" (2009b, 16). What this means is that we should run variational analysis across different instances of what is usually identified as the same practice. In other words, the same practice can have different stabilities, much like the same image, say the image of the Hallway, can be perceived differently from our firsthand perspectives. Thus, for Ihde, "In an abstract sense, all archery is the 'same' technology in which a projectile (arrow) is propelled by the tensile force of a bow and bowstring. But as we shall see, radically different practices fit differently into various contexts" (Ihde 2009, 16). Ihde discusses many examples of archery, from the English Longbow to the Mongolian Horsebow to the Chinese Artillery Bow. He discusses each of them according to rich differences. He writes:

> So what we see again is another stability in which the actual materiality of the bow, the bodily technique of use, and the cultural-historical role this technology plays as a variant. I am not claiming here to have exhausted the variations, but these three are enough to show that the phenomenological variations that now include considerations of the materiality of the technologies, the bodily techniques of use, and cultural context of the practice are all taken into account and demonstrate again the importance of variational theory with its outcome in multistability, the role of embodiment, now in trained practice, and the appearance of differently structured lifeworlds relative to historical cultures and environments. (Ihde 2009, 18–19)

The apodicticity involved in practical multistability concerns the fact that each of these practices has its own independence, and is not an inferior or developmental version of the others. There is no development continuum. Each practice is situated, robust, and requires training and expertise. In imaginative multistability, the analogue is the apodicticity of the stable pattern experienced in the perception. Yet Ihde refers to the kind of embodiment in these examples as "strong." I want to explore why strong is the appropriate term and to what degree the two kinds of multistability really are analogous.

The strong embodiment of this example is the idea that to understand multistability we have to understand an entire practice with everything that practices entail. One stability is a particular way in which a technology, like a bow and arrow, is embodied by users—not postphenomenologists. In imaginative multistability, either *we*, in the first person, are the observers and can test multistability against our actually perceiving it. In practical multistability, there are limits to the degree to which we can understand the stabilities—precisely because the stabilities are rooted in bygone or radically different lifeworlds. That is to say, I might be a sport archer today, and I might understand that there are many stabilities of being an archer, but I cannot embody these instances in my lifetime because they require expertise embedded in a socio-cultural context that I cannot expect to replicate. Here, strong embodiment suggests that, as scholars, we include more rigorous treatments of the lifeworld, describing the entanglement in substantial detail. At the same time, we lose our firsthand testability because the varied practices are ones that we cannot embody at the level of specificity at which we have described them.

By my lights, strong embodiment turns out to be slightly different than just the opposite term of weak embodiment. First, strongly embodying a practice suggests that one is competent in that practice in some lifeworld. Testing the multistability of that practice would mean that I become aware that there is more than one stable way for the practice to manifest. But I am limited in my ability to test for multistability by the constraints of what I can reasonably expect to become competent in during my lifetime. My firsthand experience of the different stabilities is less under my control. Instead, I can imagine that the practice must be multistable, and can perform thought experiments of the possible stabilities that are not grounded in the firsthand experience of being competent in those stable practices.

Second, strong embodiment suggests that someone researching the multistability of a practice is either doing so from the perspective of being competent in one or several of the stable practices or in none of them. A study, for example, of computer use probably involves a researcher who is competent in computer use in a certain context. But a study by a philosopher on, say, scientific imaging is one where the philosopher is not competent at all in the operation and interpretation of the imaging instruments. Or a philosopher can do a study of the use of a technology in another country to which she or he has not even been. In other cases, a philosopher may ponder the moral significance of the practical multistability of a technology without being able to experience firsthand what others in that moral universe actually experience.

Practical multistability, then, is a different kind of investigation than imaginative multistability. The different stabilities are not all testable by firsthand experience alone. The ascent is not experienced firsthand. Rather, the ascent is an act of "moral imagination," or the ability to project possibilities of human-technology entanglements. Ascent does not arise from the same source, firsthand experience only, in practical multistability, but from different sources. The value of practical multistability concerns whatever benefits we get from comparing our firsthand experiences to articulations of human technology entanglements that are not in our firsthand experience. If this is the case, then does practical multistability lose the firsthand rigor that imaginative multistability has? I turn to this question in the following section.

PIVOTING

Practical multistability is most often used in recent postphenomenological research, where variational analyses are run across a set of examples. Consider some of Robert Rosenberger's work on multistability and technologies. Rosenberger writes that "A centerpiece of Ihde's philosophy of technology is his claim that a technology can be embodied in various ways for various purposes. He calls this *multistability*" (Rosenberger 2009, 175; Ihde 1990, 144). Rosenberger goes on to state that, "Though we may typically embody a technology in a specific way, there are always other stable and coherent ways that this relation can occur" (Rosenberger 2009, 175).[2]

Rosenberger wants to use multistability as a descriptive framework for human practices. In one essay his goal is to emphasize "the importance of factors such as bodily habit and phenomenological space for computer design projects, and for more general conceptions of technology" (Rosenberger 2009, 179). He describes multistability with an example of a glass bottle:

> For instance, a typical glass bottle can be used for purposes other than holding liquid. It can be used as a vase for a short-stemmed flower. It can be used as a base for launching "bottle rocket" fireworks. A bottle can be broken and used as a sharp weapon. One can blow air across the tops of bottles containing different amounts of liquid to produce resonating notes. Other uses are possible. As multistable, the technology supports a number of coherent embodied relations, or stabilities. (Rosenberger 2009, 175–176).

Notice that Rosenberger's description has a different—what I would call—pivot point than Ihde's archery example. Ihde pivots on the practice of archery. Rosenberger's description pivots on the typical glass bottle. By pivot, I mean *that* which allows the variation to make sense as a variation. For Ihde and Rosenberger, the variations make sense because something remains constant in the variation. Ihde wants us to know that the practice of archery should be kept in mind as remaining constant, despite its variations. Rosenberger wants us to know that he is talking about a glass bottle as an artifact that should remain constant as we vary the types of ways it can be used in practice. When he shifts to computers in the same article, he pivots on the functionality of computer in the computer-user entanglement (e.g., the pattern of stability of a functioning computer versus the pattern of stability of a malfunctioning computer). The pivot is the degree of function of the computer hardware or software.

This difference refers to a subtlety of multistability that has to do with how it is used as the basis for descriptive research. Postphenomenologists highlight pivot points on which to run variations. The typical glass bottle could also be varied according to one of its practices, where the practice becomes subject to different variations with different materials, some of which would not involve a glass bottle. Notice that none of this is inconsistent between Ihde and Rosenberger; both use pivots as ways of framing their subject matter and running meaningful variations.

This difference in pivots sets up different uses for practical multistability. Consider Rosenberger again. Multistability, for him in the same paper, has to do with "what it means for a person to relate to a technology in terms of one stability rather than another. To embody a technology, one must conceive of it in certain ways, must possess certain intentions, and must comport one's body in a certain manner" (2009, 176). He uses the term "relational strategy" to refer to the particular configuration of bodily habits, intentions, and conceptions that make it possible for a person to take up a particular stable relation. His example is the magnifying glass. His pivoting on the technological object shows that there are different stabilities for using a magnifying glass such as magnifying something or starting a fire. In order to embody the stability, users must gain experience with the magnifying glass, develop a posture toward it, and practice. The relational strategy then may have started with an initial posture and moved into a skilled one, even though the entanglement is still a particular stability (2009, 176).

Pivoting on an ordinary artifact and not on a practice sets the investigation off in various directions. It depends on the choices that the postphenomenological describer must make as part of his or her chosen method. To pivot on the artifact requires that we assume the identity of the artifact remains constant across the variations. This pivot stresses the degree to which the material of the artifact and human intentions can create different uses. To pivot on the practice requires that the identity of the practice remains stable. This pivot stresses the importance of social, cultural and historical context in shaping what people do with technologies. Each pivot requires that we assume that something remains constant and that the reader will not question the identity.

Peter-Paul Verbeek makes a different choice, as I read him. In *What Things Do*. Verbeek includes a discussion of multistability. One of Verbeek's central arguments is that technologies should be understood in *their activity* in contrast to the assumption that technologies are no more than the values that humans deliberately put into them or are the sources of technical values that can overdetermine humans. Part of understanding technologies as active contributors to the world is to understand them as not having intrinsic properties. He writes that "one pitfall that needs to be avoided in this analysis of the ability of artifacts to co-shape the relation between human beings and world: this ability must not be conceived as an intrinsic property of the artifact itself" (Verbeek 2005, 117). He claims that even technological scripts, such as the speed bump's imperative to get us to stop, are not ingrained in the technologies themselves. Technology is not something that can be referred to independently of the particular situation in which humans engage with it. He quotes Ihde's idea that "were technologies merely objects totally divorced from human praxis, there would be so much 'junk' lying about. Once taken into praxis one can speak not of technologies 'in themselves' but as the active relation pair, human technology" (Verbeek 2005, 117).

Verbeek understands multistability in a sense that can be nuanced, though, from Rosenberger and Ihde, despite certain affinities. It is not only the idea that technology can be taken up in different ways, but that the technology's identity at any one point in time is entirely derivative of that user entanglement. It is a mistake to pivot on a technological artifact or practice because doing so assumes an identity of the technology that can be separated from the human relation. For Verbeek, multistability is the difference of identity between two examples of human-technology entanglements. Though he is interested in difference, he finds a pivot point in the standards used for technology development. Consider Verbeek's research on expecting parents, unborn children, and obstetric ultrasound technology. He wants to show that the technology entangles humans in such a way as to transform how societies engage in moral discourses. He write that ". . . the very possibility to have sonograms made at all, and therefore to detect congenital defects before birth, irreversibly changes the character of what used to be called "expecting a child." (Verbeek 2008, 16). The fetus becomes imagined in such a way as to confer an identity on it that changes our moral intuitions about the status of the fetus and the abortion controversy. Instead of being an unseen phenomena, the fetus is brought forth in ways that accord person-like qualities to it that viewers can readily identify with. These person-like qualities are not accidental, but are based in part on how the technology is designed so as to produce such an image. His pivot in this case is the sheer difference of the standards that are chosen to imagine the fetus. The standards include the size of the screen, the size of the fetus on the screen, the "realism of the image," the software used to detect risks, etc. (Verbeek 2008, 16). Each of these is part of a cluster of technical standards that go into the image produced by the sonogram. The standards themselves, then, are constitutive of the image. Verbeek's argument could be constructed as claiming that the current set of technical standards used in sonograms produce a fetus that has person-like qualities. A different set of standards, or no such standards at all (i.e., before sonograms existed), would produce a different image, and have a different impact on people's moral intuitions. Standards bring into being the stabilities of the human-technology entanglement. Standards also are mediated by the social and cultural context of those who create them, an insight consistent with other science and technology studies research on standards (Busch 2011; Busch and Bingen 2006; Busch et al. 1994).

Evan Selinger, in his study of the Grameen phone, has an even different pivot. The Grameen phone is the well-known example of the micro-credit loans to women to purchase cellular phones that they could use to conduct business (Sullivan 2007). Selinger is concerned with whether the Grameen phone is being assessed in ways that actually explain whether it has encouraged economic development, and wonders about this especially in the case of quantitative forms of assessment such as survey studies and neoclassical economic theories. He asks, "What aspects of practice are occluded when quantified analysis . . . only foreground selective consequences of participating in VP, notably the 'satisfied preferences' that can be detected from a 'bird's eye' perspective?" (Selinger 2008, 31). For Selinger, the human-technology entanglement of the Grameen phone takes on a stability depending on the form of assessment or analysis used to understand its impacts on economic development. Selinger's pivot is the perspective from which to view the entanglement, and he primarily varies the "quantitative perspective" with a "qualitative anthropological" perspective that "does not disembody or disembed the subject it studies" (Selinger 2008, 31). Based on this variation, he finds reason to worry about the success of the Grameen phone.

Ihde, Rosenberger, Verbeek, and Selinger help to illustrate a key component of the theory of multistability that I want to defend. Multistability is of two kinds, imaginative and practical multistability. Most postphenomenological research is on practical multistability. Practical multistability is describable because the philosopher has chosen a particular pivot point from which to vary the technology being studied. Practical multistability is always a matter of pivoting. The issue I want to raise here is that Ihde, Rosenberger, Verbeek and Selinger all use different pivots when they apply the concept of multistability. For each philosopher, which pivot to use was a choice that was made, in some sense, when they set out to do their projects. What informed these choices? Are certain pivots more conducive to some projects than others? And is there any reason to choose a pivot beyond its being a rhetorical move that better situates the point that the author is ultimately trying to make? It is these questions on which the future of postphenomenological research rests. Can these choices be defended?

WHAT IS THE VALUE OF POSTPHENOMENOLOGY?

Postphenomenological research like that of Rosenberger, Verbeek, and Selinger[3] does not aim only at philosophers. Each of them aims at communities outside of philosophy. They all especially want designers to consider the moral and practical implications of their designs in terms of the human-technology entanglements that come away from them. Selinger does not aim only at designers, but at development ethicists who are satisfied with quantitative analysis. Verbeek also seeks to address biomedical and clinical ethicists who may underestimate the importance of technology. But for client communities like designers to take this work seriously, postphenomenologists will have to defend their choices of pivot points to those who will wonder whether the description of variations should influence what they do or whether it is only the ploy of someone trying to harness the power of rhetoric to assert their views.

To accomplish this, there needs to be a larger discussion on what possible defenses of pivots may be, and that is not to say that there needs to be one defense for all. But there needs to be caretaken. Based on this view, the continued success of postphenomenology depends on whether practitioners can (1) develop transparent criteria and concepts that can be used and interpreted across multiple interactions and (2) develop an interactional ability with at least one client community. The transparent criteria would serve to demarcate particular methods of postphenomenological analysis that can be expected to or aim at certain results that are useful for client communities. Transparent criteria are offered for activity theory, which draws on ethnomethodology, in the case of the client communities of computer interface designers (Kaptelinin and Nardi 2009). I think activity theory could serve as an example for how postphenomenologists can develop transparent criteria for the method behind what they do and its value to client communities. Interactional ability refers to a basic understanding of a different domain of work or discipline and some competency or fluency in the language of its practitioners. These skills would allow the philosophy to communicate with practitioners and have a sense of how postphenomenology can contribute to improving their work.

One starting point for developing transparent criteria begins with the coverage of multistability I have given in this chapter. If postphenomenology is defined in terms of multistability, I can identify several stages that occur in all of the studies I explored. First, it needs to be assumed that there is multistability of practices. Second, the researcher must select a particular case. Third, the researcher selects and justifies a pivot point. Fourth, the researcher varies the case along the pivot point. Fifth, the variation creates room for discussion of the implications of the different variations. Future work needs to chart out these different steps to be convincing to others who may use the studies for their own purposes. From that, more coherent, defensible criteria could be developed. In terms of interactive ability, postphenomenologists should explore literatures on how to establish "trading zones" (Gorman 2005, 2002; Gorman and Mehalik 2002) with practitioners in other disciplines. Trading zones often occur when people in different disciplines and domains collaborate effectively using a shared language that they have developed through their interactions with one another. Postphenomenologists who develop these languages with practitioners in other domains will be in a better position to do research on the multistability of the technologies in those domains that will be useful to the client communities.

NOTES

1. Personal correspondence.
2. Here, practical multistability is a more fine-grained elaboration of the familiar ideas that technologies are polyvalent and polymorphic. That is, technologies can serve more than one set of values and can be used for purposes other than those that were intended by the designers.
3. Ihde is intentionally excluded.

REFERENCES

Busch, L. (2011). *Standards: Recipes for Reality*. Cambridge, MA: MIT Press.

Busch, L., & Bingen, J. (2006). Introduction: A New World of Standards. In J. Bingen, & L. Busch (Eds.), *Agricultural Standards: The Shape of the Global Food and Fiber System* (pp. 3–28). Dordrecht: Springer.

Busch, L., Gunter, V., Mentele, T., Tachikawa, M., & Tanaka, K. (1994). Socializing Nature: Technoscience and the Transformation of Rapeseed into Canola. *Crop Science, 34*(3), 607–614.

Cerbone, D. (2009). (Book 1) Ironic Technics; (Book 2) Postphenomenology and Technoscience: The Peking University Lectures. [Book Review]. *Notre Dame Philosophical Reviews*.

Forss, A. (2011). Cells and the (Imaginary) Patient: the multistable practitioner-technology-cell interface in the cytology laboratory. *Medicine, Healthcare and Philosophy, Forthcoming*.

Gorman, M. E. (2002). Levels of Expertise and Trading Zones: A Framework for Multidisciplinary Collaboration. *Social Studies of Science, 32*(5–6), 933–938, doi:10.1177/030631270203200511.

Gorman, M. E. (2005). Levels of Expertise and Trading Zones: Combining Cognitive and Social Approaches to Technology Studies. In M. E. Gorman, R. D. Tweney, D. C. Gooding, & A. P. Kincannon (Eds.), *Scientific and Technological Thinking* (pp. 287–302). Mahwah, NJ: Lawrence Erlbaum Associations, Inc.

Gorman, M. E., & Mehalik, M. M. (2002). Turning Good into Gold: A Comparative Study of Two Environmental Invention Networks. *Science, Technology and Human Values, 27*(4), 499–529.

Ihde, D. (1976). *Listening and voice: a phenomenology of sound*. Athens: Ohio University Press.

Ihde, D. (1977). *Experimental phenomenology: an introduction*. New York: Putnam.

Ihde, D. (1990). *Technology and the lifeworld: from garden to earth* (The Indiana series in the philosophy of technology). Bloomington: Indiana University Press.

Ihde, D. (1998). *Expanding hermeneutics: visualism in science* (Northwestern University studies in phenomenology and existential philosophy). Evanston, Ill.: Northwestern University Press.

Ihde, D. (2009). *Postphenomenology and technoscience: the Peking University Lectures*. Albany: SUNY Press.

Kaptelinin, V., & Nardi, B. A. (2009). *Acting with Technology: Activity Theory and Interaction Design*. Cambridge, MA: MIT Press.

Riis, S. (2008). The Symmetry Between Bruno Latour and Martin Heidegger. *Social Studies of Science, 38*(2), 285–301, doi:10.1177/0306312707081379.

Rosenberger, R. (2009). The Sudden Experience of the Computer. *AI and Society, 24*, 173–180.

Selinger, E. (2008). "Does Microcredit Empower? Reflections on the Grameen Bank Debate". *Human Studies, 31*, 27–41.

Sullivan, N. P. (2007). *You can hear me now : how microloans and cell phones are connecting the world's poor to the global economy* (1st ed.). San Francisco: Jossey-Bass.

Verbeek, P.-P. (2005). *What Things Do: Philosophical Reflections on Technology, Agency, and Design*. State College, PA: Pennsylvania State University Press.

Verbeek, P.-P. (2008). Obstetric Ultrasound and the Technological Mediation of Morality: A Postphenomenological Analysis. *Human Studies, 31*, 11–26.

Verbeek, P.-P. (2009). Ambient Intelligence and Persuasive Technology: The Blurring Boundaries Between Human and Technology. *NanoEthics, 3*(3), 231–242.

Chapter 6

Hospital Architecture and Design in Postphenomenological Perspective

Lars Botin

This chapter will deal with the dialectics of body and life in an overall phenomenological/postphenomenological and socio-technical reading. This will hopefully lead in a direction where appropriate action in the field of health and hospital design is made possible. It is the assumption that health is what falls in between body and life. Health is personal and general, measureable and immense, and as such it is a hybrid (or chimera) constituted by apparently polar entities.

In the movie *Dead Again* (1991) directed by Kenneth Branagh and starring Andy Garcia the opening-scene depicts an old and sick journalist. He is interviewed and during the talk he asks the interviewer for a cigarette. He has a tracheotomy in the throat and his breathing is obviously seriously disturbed. As he smokes the cigarette through the hole in the throat the camera zooms on the delighted face of the smoker. There is a strong sense of fulfillment and deep satisfaction, and at the same time the viewer is left with a sense of horror and despite. The action of the smoker is certainly unhealthy (seen in a physical perspective), but in the moment and situation we learn how the person on the screen feel happy and thrived—satisfying needs and wishes that are linked to his personal life and identity. The smoke gives him the power and will to talk about his life, and then the story begins.

What do we learn from this account, which is certainly not unique? We learn that the human being has a pragmatic (perhaps even irrational) relation to the physical body that sometimes leads to harmful actions in relation to the very same body. We learn that our concept of body and life is variegated and personal, and health is both physical and also related to well-being. The figure in the movie feels well the moment he smokes. Well-being, gratification, satisfaction and more constitute, along with sane and normal lungs, livers, kidneys, hearts and so forth, our health. Society has for decades been focusing on health as related to the measurable body and on good physical health as being synonymous to good life. Here we have the clash, because we learn from the movie that situate good life is not always equal to physical healthy action. Science and technological development has been following the path of society (or vice versa) on viewing the physical health as synonymous to good life, and the longer we live the better.

Postphenomenology has a similar ambivalent and hybrid understanding of life and health, and how we enact our lives through engagement and interaction with our surroundings mainly mediated by and through technology (Verbeek 2005, 2011). The concepts of multistability, material hermeneutics and lifeworld are all pointing toward the importance of subjects' different appropriations, uses and understandings of technology (Ihde 1990). In this chapter I will discuss a specific technology, hospital architecture, in a hybrid and multistable perspective, where I will discuss the dynamics and emergence of life and body in relation to the apparently fairly fixed framework of architecture in an institutional setting, e.g., the hospital.

Hospital architecture is generally considered as mere containers of medical functions with no noticeable remarks on form and architectural quality. In this particular perspective I shall try to see hospital architecture as health technology in a postphenomenological perspective, because it opens up for new understandings of how architecture is shaped by users and users are shaped by architecture. These co-constitutions are variegated and multiple interdependent with the situational use and meaning construction that takes place in the human-technology relations and interactions (Ihde 1990, Verbeek 2005, 2011).

A search in major theoretical and historical contributions to the understanding of architecture shows a surprising lack of samples from the realm of hospital architecture and construction. Historically, at least in the twentieth century, there is very little evidence of architects paying particular attention toward hospitals and similar institutions. Whereas schools, libraries, churches, concert halls, museums, governmental and municipal buildings, cemeteries and even power plants have had the attention of both commissioners and architects then hospitals, on a general level, has been forgotten or neglected in architecture and has remained buildings and constructions which were thought of as *functional containers* for work-procedures and practices of medical staff and for temporal permanence of patients.

In the following paragraphs we will look into the historical groundings for hospital architecture in a Danish context in order to trace the genealogical reasons of the current models. We will also discuss the potentials of phenomenology, and postphenomenology, when it comes to frame what can be made in order to think meaningful and appropriate design for different types of users, might they be medical staff, administration, patients and/or relatives. Central in the discussion on the container hospital of modernism, is the assumption that this typology is the ontological expression of modern evidence-based medicine, hence difficult to escape or disregard when it comes to designing for a lifeworld that is still dominated by the ideals and models of modern Western man.

THEORETICAL BACKGROUND

One of the most evidential aspects of our existence is death and the American phenomenologist Drew Leder has drawn attention to how medical practice, based on evidence, envisions death as the measurer of life, which means that the statements and representations of evidence take on a dead and nonliving guise. Leder writes that medical education is initiated with dissection of corpses and a clinical case often ends up in the laboratory of the pathologist (Leder 1992:22). From the beginning to the end of almost every practice within the hospital system, we see that the dead body or the representation of the dead body is what evidence is extracted from. Leder has made the observation that as we are examined in the clinic: "The patient is asked to assume a corpse-like pose, flat, passive, naked and mute. The entire ritual and context serves to reduce the living body to something almost dead" (Leder 1992:22).

The statements of evidence which is produced within the system and which are clear, unmistakable, unquestionable and beyond dispute are based on extracted data from either dead bodies, or immobile and mute representations of death. It seems obvious that such statements could and should be questioned in the perspective of the lived body, because the evidence created only takes into consideration our physical body as matter.

The French philosopher Michel Foucault has dealt extensively with the problem of sight and the image, hence evidence; in medical practice in *The Birth of the Clinic* (1963) and especially the chapter concerning "The visible and the invisible" is eye-opening in this regard. Modern medicine has since its emergence overly focused on the eye as revelatory of truth: "Truth, which, by right of nature, is made for the eye . . ." (Foucault 1963/2003:204) and in a frenetic and ongoing process toward revelation something happened in modern medicine, which through the employment of sophisticated technologies replaced life as bearer of existential and corporeal truth. Foucault writes: "That which hides and envelops, the curtain of night over truth is paradoxically, life; and death, on the contrary, opens up to light of day the dead coffer of the body: obscure life and limpid death . . ." (Foucault 1963/2003:204).

Since Foucault wrote *The Birth of the Clinic* technological development and innovation has in many respects replaced the meaning and content of death as a revealer of truth. Dissection has become less important in order to gain knowledge about disease, because the vigilant eye of the dissector has been replaced by micro-technologies that can be introduced into the living body in order to detect and map the disease. Life is lived, monitored and registered during lifetime through technologies and stories which are told about life as life goes on. Death was, in the past, a distant, static and objective observer of life and disease, whereas now technologies have become an entangled and dynamic agent in the intrinsic relationship between life, death and disease. Distance has gradually disappeared and been replaced by a perception of life, time and space where everything has become contemporary, thanks to the capacities and potentials of technology.

The ability of contemporary technologies to look into the most secret places of our bodies and cast light upon these obscure matters has replaced death. We might say that the virtual image and representation of our body has replaced the carcass and rendered death insignificant to anybody than the individual and relatives. Our living bodies have in this sense obtained a certain kind of immortality and death has become antonymous to the body.

EMBODIMENT

French philosopher Maurice Merleau-Ponty introduces the reader to the concept of *body skhema* in one of the major works of phenomenology: *Phenomenology of Perception* (1945) wherein he emphasizes the importance of body as something more than a conglomeration of parts and bits.

In my reading of the concept it becomes a corporeal attitude, based on ethical and aesthetic considerations. This reading goes beyond the purely spatial meaning of aesthetics, that is, sensing and experiencing, where relationships of our bodies with the outer-world are seen as either positional or situational, hence something that can be documented and described in a rather objective and schematic way, according to the conventional use of the term scheme. It implies that our bodies are entangled and involved with other bodies and things in the world, and we create corporeal meaning along the way, constantly becoming anew (Merleau-Ponty 1945/1994:42–44).

The *body-skhema* as attitude is a pre-noetical factor for creating a fluid and dynamic identity that comes before social ordering and construction. It is layered in our bodies, which are not entities detached from the mind, but determinant for conception and understanding. This means that our bodies have a capacity and potential that is innate (pre-noetical) and that it is activated as we are born into this world. The capacities and potentials are stimulated and enhanced by the subjects and objects of our surroundings and our bodies become part of an ongoing process of metamorphosis and change. The basis of this everlasting social and technological construction of our bodies is our *corporeal attitude* toward stimuli, events, accidents, experiences and situations and this attitude (*skhema*) has both an aesthetic and ethical character.

Our bodies are in constant movement and according to Merleau-Ponty we create from the very beginning of our existence a scheme for appropriation of our surroundings; we construct an attitude, a mode of perception, which is common to every human being although embodied within the individual subject. We seek patterns and try to order our presence in space and time, in blocks of past, presence and future. Entities of time are interrelated, overlapping and in the end make it so that we act and interact as human beings in both aesthetic and ethical ways. All of this occurs or becomes due to our *body skhema*, which is generated, in the single individual; nevertheless, it is not subjective or unique because it is related and based upon bits and parts of our bodies that are common to every (normal) human being. And it goes the other way around, as well, which means that aesthetics, that is, the way we perceive the world through our senses and bodies, is general and provides common meanings and understandings that can be structured, schematized and ordered, hence becoming similar to scientific statements according to the traditions of Western science. The American pragmatist philosopher John Dewey, in *Art as Experience* (1934), states that aesthetics ought to be placed in the core of everything, because it is the way our bodies relate to and understand the world. The German philosopher Hans Georg Gadamer was in agreement with Dewey about the potential of art as an epistemologically sound and valid way of understanding the world and reality, when he wrote: "In the experience of art there is present a fullness of meaning which belongs not only to its particular content or object but rather stands for the meaningful whole life" (Gadamer 1960/1992:63).

The *body skhema* is, as I see it, the mediator, which can be summarized in the following list:

- General quality and capacity of the human body (aesthetics).
- Common attitude and perception of the body (ethics).
- Universal and cyclical perception of time and space, hence fusion of past, present and future (experienced physics).

Beside the fact of our bodies being schematic they constantly, in their number, volume and variety, confront and test the limits and borders of existence. And here as well it works the other way around: we, as bodies, are constantly tested and confronted by the limits and constraints of context.

The American philosopher of technology Don Ihde writes extensively about our dialectic experience of our bodies and names the two our *here-body* and our *there-body*. Our *here-body* is the body, which is seeing, touching, experiencing and imagining in first-person, and our *there-body* is observing from a distance the phenomena of the world and imagining in second-person. We can do both, but we are inclined to live and imagine in either an active or a reflective manner. Ihde brings forth the sample of how students were asked to imagine the act of parachuting and describe what they saw and felt. Some of the students would describe how they felt the wind, the speed and the "sense of vertigo in the stomach" (Ihde 2002:4). Others would describe how they saw an airplane and the opening of a door and the jump of a figure. We are both *here-* and *there-bodies* and the only way we can fully perceive and understand the world and ourselves is by being aware of this condition engaging in a dialectics of experience and imagining. Technologies can be constructed in order to take part and enhance this dialectic condition and become more than just a tool or machine that replaces human physical or/ and mental effort.

It has been stated several times that technological design and development, as engineering science, is relying on the natural sciences and the principles of value-free and objective knowledge production. This means that major part of technological development engages with the body on a distanced level, hence referring solely to our *there-body*. We explain by observation how a situation or event evolves in time and space, and in the end we try to deduce some kind of meaning from the observation. This is a way of the body that remembers, recalls, and imagines; and a central way of creating knowledge that is useful in both everyday life and in science. But science and engineering tend to overlook the presence and importance of our *here-body*. This is the body that enacts, engages and experiences in time and space, and the learning from this direct situational and positional involvement is missing from most scientific or applied scientific work, and tends to be regarded as subjective, casual and out of control. If we regard our *here-bodies* as wild, savage, primitive and/or naïve, which is how science and applied science have tended to do for centuries then there will be no possibility for an intimate and experiential reference in the realm of our *there-body* universe of reflection and imagination.

Ihde's dialectics of embodiment where the *here-body* and the *there-body* are simultaneously present in the reflexive temporal and spatial experience can open up for a new understanding of hospital architecture as technology we embody through use and practice. Architecture is seldom considered as technology in line with hammers, pumps or computers, but architecture is technology with specified functions and uses that are allegedly concerned with the restoration of health, alleviation of pain and comforting the distressed, which are all *here-body* experiential. Nevertheless it seems as if the modern and late-modern hospital architecture focuses on the *there-body* making way for body counts and abstract body images in order to optimize in relation to economic efficiency.

Ihde is mainly concerned with the "multistable ambiguities" of the two body perceptions in relation to virtual reality, but in this case I think that these "ambiguities" could as well be identified and operationalized in relation to analogous hospital architecture, perhaps in search for a different type of virtuality, for example, virtues of the two body perceptions. Ihde writes: "Combining these multistable ambiguities, one can begin to appreciate how complex the issues of virtuality may become. The opening to a sliding perspective from the multidimensional experience of my here-body toward the image-body perspectives lie within these ambiguities" (Ihde 2002:6). Both bodies are emergently and full-blown present in a hospital context and there should be an "opening to a sliding perspective" in order for the two bodies to interact meaningfully in and with the technology of architecture.

Another way of addressing the dialectics of embodiment and technology is present in what has been coined as *hybrid imaginary* (Jamison, Christensen and Botin 2011) because a perspective which is constructed out of various components. "A hybrid imagination can be defined as the combination of a scientific-technical problem-solving competence with an understanding of the problems that need to be solved. It is a mixing of scientific knowledge and technical skills with what might be termed as cultural empathy, that is, an interest in reflecting on the cultural implications of science and technology in general and one's own contribution as a scientist and engineer (architect), in particular. It can be thought of as an attitude of humility or modesty, as opposed to arrogance and hubris, in regard to scientific and technological development and, for that matter, to any kind of human activity. As such, a hybrid imagination is often manifested collectively, involving collaboration between two or more people even when it is not explicitly part of a social or cultural movement" (Jamison, Christensen and Botin 2011:4). This approach is needed in order to deal with various types of challenges of contemporary reality, and in this specific perspective hospital architecture, addressing the sustainability challenge, the societal challenge and the techno-scientific challenge, which according to the hybrid imagination should be met with engagement, cross-disciplinarity and contextualization (Jamison, Hyldgaard Christensen and Botin 2011:7). Hybrid imagination is hence about both the embodied and the disembodied perception of technology in the perspective of Ihde.

Integral part of genealogical, postphenomenological and hybrid imaginary investigations is an in depth understanding of the historical context out of which the phenomenon, in this case hospital architecture, has become. In the following I shall address the historical evolution of hospitals in a Danish context in order to show how modern medicine constituted in de-contextualized disciplinary monotypes of technicality and furthermore address the inappropriateness of these types of structures in relation to our multistable bodies and *hybrid imaginary* as described in this paragraph.

A BRIEF HISTORY OF HOSPITAL ARCHITECTURE IN DENMARK

If we look at the structure of an architectural design, it consists of various rooms and spaces, where a certain kind of hierarchy is at hand. Each room has its own qualities and functions, different sizes and users. Some of the spaces are accessible for everyone, whereas others are more exclusive. Some spaces are accessible for everyone at all times, whereas others are accessible for everyone, but only one at a time. This is the way we tend to live our analogous architectures in everyday life.

The first hospital in Denmark, Frederiks Hospital, was constructed in 1751–1758 by the most renowned architect of the period, Nicolai Eigtved. Eigtved, who was commissioned by the king Frederik V and chancellor A. G. Moltke to design and construct a hospital for the poor and diseased, as part of the royal plan for the new Frederiksstad, where the king himself resided. The architecture is fairly classical and inspired by Italian and French late baroque palace building. The construction does not reveal function, but merely abides to stylistics and aesthetics of the planned buildings in the context.

The interesting story about this first building of the Danish national health care system is the social and procedural aspects of the construction, where Eigtved during the first years of the construction kept to drawings and sketches that he had put up together with Moltke and the Frederik. By 1753 doctors began to interfere, requesting a basement under the building because of the moisture and dampness of the ground. Furthermore they requested the building to be a height of 3.7 meters in the rooms and that the central courtyard be furnished to accommodate the recreation of the patients. Eigtved was very reluctant to meet the requests of the doctors and an agreement between the architect and the doctors was never reached, as Eigtved died shortly after (1754).

Today the buildings are housing the National Arts and Crafts Museum, which clearly underscores the general quality of the architecture, but also indicates the absolute detachment from specific use and function.

The process of these central years in the construction show that the architect did not have any real insight or interest in the affairs and procedures of the hospital as the doctors had no insight or interest in the affairs of the architect.

The reasonable requests made by the professional staff of the hospital were never really fulfilled, but the follower of Eigtved, Lauritz de Thurah, managed to satisfy the doctors with technical fixes for ventilation and light. The only request that was met in full by the new architect and his commissioners was the recreational courtyard, which still today stands as one of the most beloved courtyards in Copenhagen.

The hospital served as a model and was the hospital of the nation for 150 years, and it was not until 1910 that a new kind of hospital was generated upon the ideas of hygiene and the knowledge upon bacterium. The form of the new hospital was inspired by separated pavilions where wind and light could penetrate space and eradicate moist and dark areas, considered as bearers and conditioners for bacterium and disease. The pavilion hospital was not new in a European context, where the first of the kind was designed and erected in the middle of the nineteenth century in Paris (Hotel Lariboisiere by M. P. Gauthier). During the following decades the model spread to all the Western societies, and as the old Frederiks Hospital was finally closed in 1910, the new National Hospital, by Martin Borch, was built on the model of the pavilion.

The pavilion has obvious advantages for what concerns the architectural, analytical aesthetic and experiential aesthetic aspects. It lends toward a docile control of the landscape and emphasizes the relationship between nature and health. On the other hand it creates a number of problems on an infrastructural level, where the singular entities are isolated and distanced, and risk becoming small enclaves of despotism, power and unreason. The pavilion structure is also unable to cope with the diseased body in its complexity, because each pavilion had its discipline and disease. The hospital got rid of the dark and crowded rooms of the old structures, but as the new structure was implemented on a national level new problems occurred, where the decentralized structure of the pavilion hospital became both solution and problem.

The National Hospital lasted for half a century, but became actually outdated in a few decades. The new ideas of optimization of time and space, which the technological development in mainly the US had evoked and prorogated showed in block hospitals as early as 1911 and soon became the model in American metropolis.

The advantages of the block hospital were obvious seen from an economic and administrative perspective, but also from the doctor's perspective. Now he/she was able to communicate with peers from other branches of medicine and surgery, hence making interdisciplinary care available to patients with multiple diseases.

Central and general technological departments like X-Ray and the pharmacy could serve the whole hospital without any problems, and mechanical infrastructural communication systems in tubes could cover the whole body of the building. It goes without saying that the costs for building, heating, plumbing and so forth were much lower than in the pavilion model. The question of light and space was not considered a fundamental problem, because in the end dependent on the economical efforts made by the commissioners. It was only a question of building big and tall enough.

The container or block hospital, which I will discuss in the following paragraph, has had a decisive impact on how we perceive and conceive subjects and objects within the frames of the container. We have structured spaces, humans and things according to the logics of the block and the container, which has its obvious advantages but also shows permeated with alienating geometrical organicism, where the individual doctor, nurse and patient succumbs on behalf of the beautiful and logical structure.

AALBORG HOSPITAL SOUTH

The Danish hospital system underwent a radical change in the 1960s due to the economic conjunctures, and a sublimation of the modern and technological rationale became possible. This resulted in an intense productivity in masonry. New hospitals were built all over the country and this count as well for the county of North Jutland, where the hospital of Aalborg (south) stand as a perfect example of the sublime model of modernity and technology. The particular model was not Danish in its outset. It was imported from Italy where the main corpse of the architectural assembly was the vertical container of departments of beds. What was to some extent stable and static was contained in the main corpse, whereas the more dynamic and temporal departments were situated in buildings and spaces attached, or detached from the main corpse.

The model was totally in line with the self-esteem of modern medicine and mirrored the intimate relationship between science, technology and medicine as it envisioned the body as a hierarchical distributed assembly of bits and parts. Hardy and Tansey write in a recent publication on the *Western Medical Tradition 1800 to 2000* (2006) that: "As symbols of modern scientific medicine, of medical progress, and of egalitarian modern social policies, hospital reached their summit of eminence in the 1950s and 1960s. This symbolic identity was now reflected not only in the building of new hospitals, but in the modernist architecture that characterized them" (Hardy and Tansey 2006:441).

If we look closer at the model as it manifested itself in Aalborg in the 1960s, and although the hospital has grown in size and importance ever since, the ground model has remained unaltered and unquestioned; we see that the main characteristics of modern Western thought concerning the scientific view of body is mimetically layered in the building.

The first thing we encounter is the main entrance consisting of an impressive lobby, where we find retailers, information-lobby, elevators and on the outskirts we have got outpatient departments. Entering the elevators we access the departments of bed wards. If we take them to the top we find the department of neurosurgery and medicine, as well as the department of intensive care. Going down the "body" we meet the department of eye, throat, heart, thorax ending up in orthopediatry.

In the cellar we find the infrastructural body-parts of the building, and as well corridors leading to the morgue. The population of the basement consists of technical and transportation staff, which also forms the lower part of the hospital hierarchy.

The content and displacement of the building mimes the body—from head to foot—which is quite logical and natural as we try to find our way through the building. It also mimes the importance of the various human body-parts where the head is situated on top in tranquility, silence and a stance of illuminate clarity is at hand due to the position on top. As we move down we get ever closer to the chaotic condition of the body. The bells of moving elevators, moving bodies often by the aid of technologies, relatives and nursing staff increases in numbers and the confusion of outside life is reflected in action and behavior.

In the lower part of the body-building the human body is considered and treated in larger chumps whereas specialization and particularization is increasing as we move upward ending in the eye and brain. The Cartesian paradigm of mind over body and appropriate analytical action and reflection is showing in an exemplary way in the hospital body-building.

We were talking of how the vertical corpse of architecture contained the stable body, whereas the more dynamic body and the fallacies of this more dynamic body are contained in buildings attached or detached from the main corpse. Cancer, lung diseases and gastro-enterological diseases are dealt with in buildings that are outside the main corpse, because dangerous and filthy parts of the human body. Much in the same way the "dangerous" technological body-parts of the hospital building are hidden in the cellar and disguised in walls and white painted tubes.

The hospital reflects the Cartesian biological/medical paradigm, where the body is seen as a machine that can be read and translated literally into architecture, masonry and technology. It is a Platonic view on the relationship between presentation (the real body) and representation (the body of the hospital), where the representation paradoxically and in a pervert reading of the relationship in between the two, overcomes the presentation, because decisive for action and behavior of the presentation within the frames of the representation.

All of this happened well before the 1960s within the hospital system, but the vertical container building is the ultimate symbol of this evolution that according to Michel Foucault began in the middle of the nineteenth century and the logics and rationale of how the modern hospital became can be summarized in the following quotation: "Here, then, (in the middle of the 19th century) emerged a new object of study: 'Man.' The target of the prison apparatus—the body—was studied with detailed precision relegated before to insects or specimens of plant-life: marked, invested, trained, evaluated, compared, and made the carrier of signs. New modes of expertise emerged and proliferated as the knowledges developed in the prison were transported to, and transformed in, hospitals, schools, factories, military campus, and insane asylums, spurring the development of Taylorism and techniques of mass production" (Alcoff 2005:220).

It happened gradually through a century and mainly under the pressure of science and technology, where the sick and battered body was envisioned as a disturbing variable within the system. Claude Bernard (1813–1878), to many the founder of physiology as science, claimed that the hospital of the day was "only the entrance to scientific medicine; they are the first field of observation where the physician enters; but the true sanctuary of medical science is the laboratory; only there he can seek explanations of life in normal and pathological states by means of experimental analysis" (Bernard 1865 in Bynum 2006:117). The direct contact with the sick patient lying in bed and the observations made in such a set-up were far too insecure and context-related in order to lay the groundings for scientific knowledge, and the doctor moved away from bedside into the laboratory where he experimented on dogs, rats, rabbits, pigs and eventually human beings. He extracted bits and parts from the human body and transferred the matter into the technological setting of the laboratory where he by the means of precise technology was able to produce sound and reproducible results. The ways and means of the laboratory gradually invaded the spaces of the hospital and in the 1950s and 1960s technical technology had finally managed to minimize the physical contact between doctors and patients and the body was engulfed by sublime technical devices for which the container structure of the hospital served superbly.

Georges Teyssot stated that the container for storing our sick bodies is characterized by a conception of the body in its *interiority* where the body is looked upon as a hierarchical congregation of body-parts and organs, where the: "autonomous organs (de)compose the whole in multiple parts, breaking up its integrity. This logic is reflected in the distribution of a hospital in various medical specialities. It is also a type of functionalism that lies at the base of all 'modernistic' architecture, which was nothing else than an application of organicism" (Teyssot 2005:73).

The social problems in tall buildings for housing shows that this organicism that spread, according to Foucault from the ontology of prison, is highly inappropriate to our physical bodies, contextual bodies and social bodies. Alcoff paraphrases Foucault on this matter in the following way: "Actions were broken down into small segments for reorganization toward greater efficiency with the aim of creating totally useful time" (Alcoff 2005:220). We actually get sick, disturbed and uncomfortable in our being as we meet the pressure of analytical and functional structure. Paradoxically this was totally ignored by all important actors in the building-boom that occurred within the Danish health care system in the 1960s, because architects, the Department of Housing, and doctors all gave into the technological logic of industrialized building (Dirkinck-Holmfeld and Heslet 2007:98).

The hospital of Aalborg South resembles a number of other examples from both a Danish and international context. Huge investments in the 1960s resulted in similar buildings in Copenhagen (The National Hospital and Herlev Hospital) and Odense (The University hospital), to mention the most important ones. We enter the building through an opening, which in many ways resembles the mouth either highly adorned like the perfectly glossed mouth of a woman or strict and sharp as the mouth of a decisive person.

DWELLING AND EMBODIMENT

Our everyday houses possess the same feature. We have got a main entrance that is often adorned by plants, small stucco sculptures and relief's and name targets. But in the major part of modern Danish housing we have got a second entrance, which leads to the utility room, and this entrance is used in everyday practice, whereas the main entrance is used on special occasions. The second entrance is placed in a less apparent position and is very seldom adorned or framed in anyway. Attached to the utility room we find the kitchen and during the last decades the kitchen has been merging with living rooms and dining rooms. The positioning of sleeping rooms and toilet(s) can vary, but generally the positioning of the main sleeping room of the adult(s) will be more isolated.

We enter the body of our private shell through an alternative entrance, which immediately puts us in contact with the vital parts of our own bodies and the body of the house: the utility room with heating and electricity main-boards and the kitchen. And although the official building has a mouth which could posit us in the very same position, then the mouth in the official building connects us to a formal, intellectual and detached mode of communication. The entrance of the private building connects us directly to internal and visceral body-parts and we fall at ease into this ambience. From there we construct our space of communication, which becomes on several scales and through different media as we move from the kitchen into the adjoining/adjacent spaces of relax and action. Radio, television, mobile phones and computers interact in a natural way with the primary and local communication between individuals. Our private homes are domains of information and communication of both local and global character.

Communication is not governed by an overall rational and static stance. It is relating to the context and spreads in the societal body. The house is how we relate to the outer world. In the words of the French philosopher of phenomenology Gaston Bachelard: "A house that has been experienced is not an inert box. Inhabitated space transcends geometrical space" (Bachelard 1958/1994:47). The house, as a piece of technology, is incorporated in a non-Euclidian manner and an integral part of how we become on every scale, time and space: "Thus the whole thing (the house) is scattered about inside me, the rooms, the stairs that descended with such ceremonious slowness, others, narrow cages that mounted in a spiral movement, in the darkness of which we advanced like the blood in our veins" (Bachelard 1958/1994:57).

The house of our memory and experience is embodied in our everyday life and the following drawings are illustrations of the house of my memory and experience. What are depicted are two houses that I lived in during my childhood and youth, hence according to Bachelard, running in my veins and totally embodied. The first house was constructed in the beginning of the 1960s, consisting of three bedrooms, a living room, a bath, a kitchen and two entrances. The second was a two-story country-home built in the middle of the 1950s, with a living room, kitchen, office, bath, hall and utility entrance, and all the private rooms on the second floor. The drawings have been made out of the embodied experience and memory of the place and are precise in neither scale nor size. As can be seen they show striking similarities: the utility entrance, kitchen and living room are closely connected. As I left for school, got home from school, went out to play and returned from soccer it always happened through the "hidden" entrance that led to the utility room and to the kitchen and living room. It never happened through the main entrance or through the entrance hall. The houses were embodied through my everyday life practice and I connected to the house and its belongings through the humble entrance of utility.

This reading of how I connect to houses of my embodied memory in the spirit of Bachelard can be enframed in a postphenomenolgical human-technology relationship, because I embody technology through which I relate to the world, that is, (I-house)—world. I have lived in thirty-five different places and it is not all of these places that I have co-constituted in this way. In most cases it has been a background relationship, and in a few cases an alterity relationship (Ihde 1990, Verbeek 2005 and 2011).

The hospital is not a house and it is not meant for dwelling, which means that we can never be true corporal beings within the framework of this kind of building. The hospital is in this sense comparable to the fire station, the power plant, the shop, the bridge, the truck and other places where people work and produce things, but there is a fundamental difference between these places and the hospital; and this has to do with their purpose. The purpose of the hospital is to try to cure and repair the vulnerable and fragile human body that has been removed from its everyday life setting; and this is fundamentally

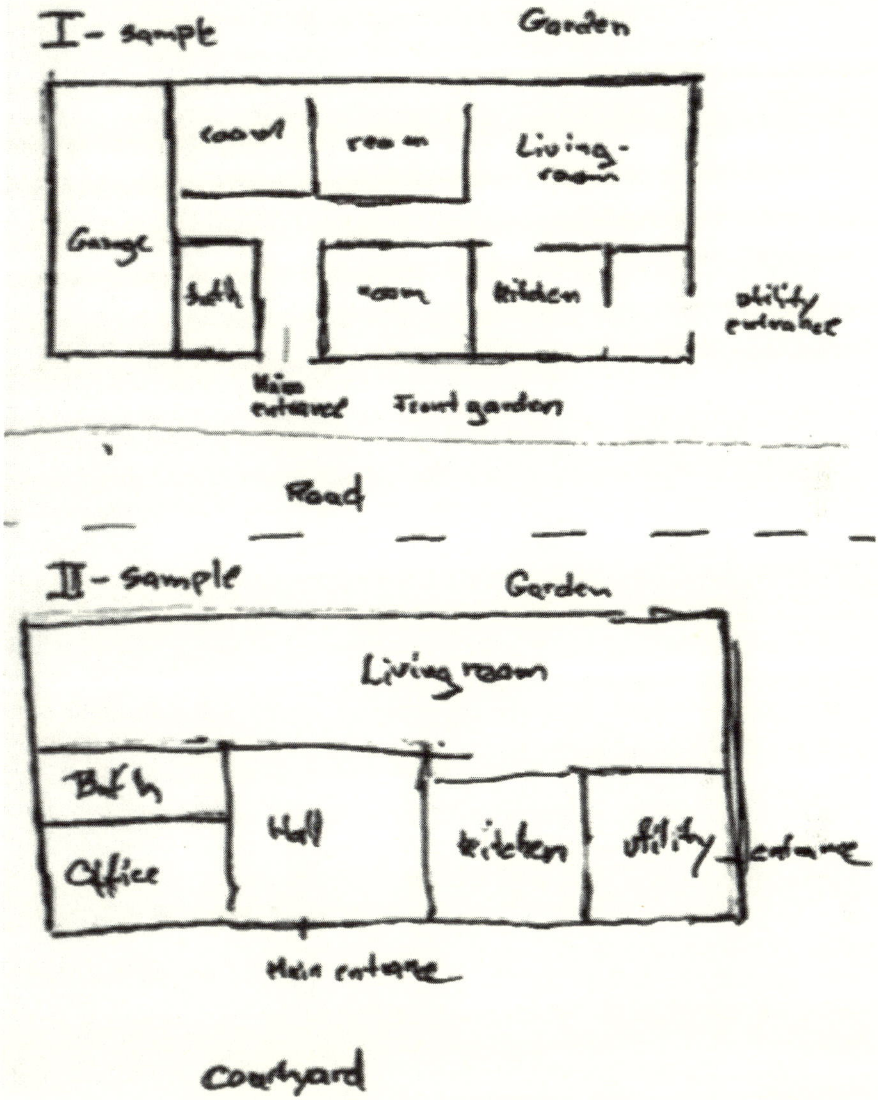

Figure 6.1 Ground plan of houses that I have lived in. Botin 2007. *Source*: Author created

different from the goings-on in a power plant or similar places, where we produce, transport, or consume goods and objects. Yet the hospital has been constructed as a place where production, transportation and consumption are at the core of everything, because our fragile bodies are envisioned as objects for repair by means of technical manipulation.

CONTEMPORARY HOSPITAL ARCHITECTURE

In this final part of the chapter I shall analyze the current situation for what concerns hospital design in a Danish context. During the last couple of years there have been several major competitions and of infusing a considerable amount of funds in the Danish health care system, $200.000 million a year for a period of ten years, in order to upgrade and improve existing structures and building new ones, can show devastating to body (on any scale) if we do not consider the inappropriateness on a fundamental level of the existing structures, because so distant and alien to how we engage with subjects and objects in our everyday lives. I shall focus on two of them in the peninsula of Jutland. In 2007 C.F. Møller Architects won the competition for The New Universityhospital Aarhus (DNU) and in the description of the project they emphasize two keywords: the metaphor of the city as a structuring and systemic device and evidence based design in relation to healing architecture where the patient is placed in the center. If we look at the plan of the project we find a spine structure of U form enframing the central building (the entrance) and meant for housing of university staff and administration. In the spine-structure are located the various clinical specializations in a way that resembles the principle of the pavilion-hospital. They are classified as quarters or sectors according to the urban metaphor.

The criterions for evidence based design are: the green hospital city, orientation and logistics, sustainability, 1 bed ward rooms, design and interior, inner climate and daylight, acoustic design, texture, artificial light, art, "the good workingplace," and IT and medico-technology. Integral part in all the criterions is the well-being of the patient and the aim is to diminish time spent within the hospital, which of course transcends the concern of well-being and introduces an economical parameter, which is not mentioned in the description of the project.

The metaphor of the city is carefully described and we get the impression that in this interpretation there are references to modernist projects like the English garden city from the beginning of the twentieth century or, in a metropolitan scale, *Unité d'Habitation* from 1946 by Le Corbusier. The aim of using the metaphor of the city as a structuring device is to create a sense of spatial familiarity, which certainly has its roots in phenomenological reading of space and things. Nevertheless there are some problems with the use of the metaphor because even though the German philosopher Hans Jonas writes that the city as societal artifact is the modern representation of the human soul in which prudence and good behavior ought to meet (Jonas 1979/1984: 30), then the city is also the overcoming and domestication of nature and a rational and instrumental tool for structuring and controlling the human body as a social entity (Hughes 2004:157 and Williams 1989/2007:47). The question is whether this dialectics in the city metaphor can be inscribed in a hybrid imaginary framework, because the ingredients are there. The concept of spatial familiarization coined with structural possibility of controlling social and individual behavior potentially constitutes hybridization, but only if this aims for a collective improvement of things. The architectural focus on the "brain" in the tall and chunked (100 meters) central building and the "scattered" body in sectors together with the early modernist references do not point in that direction. On the contrary there is a continuation of Cartesian paradigm of "mind over body" and a celebration of disciplinary specialization that prevents cross-fertilization and common engagement.

The evidence-based design for healing architecture and the criterions set for this approach could also be inscribed in a hybrid imaginary framework, because covering both qualitative and quantitative aspects of enhancement and optimization of the human condition. Nevertheless there is in the program an overwhelming and inappropriate focus upon the individual, which in many cases reflects the neoliberal narcissism and egocentrism of the New Millennium. In many ways it is phenomenology run astray, forgetting the *skhema* of Merleau-Ponty. In the program the body is purely envisioned as an aesthetical and biological/physical subject, neglecting the social and moral body of the patient.

If we turn our attention to a more recent competition in Northern Jutland. The Aalborg University Hospital (2012) then we find very similar design in the 6 proposals. All of them work with spine-structure, which in these projects is stretched in order to accommodate and work with the topography of the landscape. The metaphor of the city or part of the city (the *rambla*) is present in 5 of the proposals, hence with similar possibilities, potentials and problems as DNU.

It is also a common trait that the plane-metric motives in the spine-structures have striking similarities to the Zwastika leit-motif of the Bauhaus school from 1926, which lends a powerful dynamics to the geometrical structures, and at the same time in their iconic reference points back to rationalism of modern Western society. In the assessment of the proposals we only find the patient mentioned once in relation to the only evidence based design proposal by C.F. Møller Architects, whereas the keywords for assessment are: functionality, architecture and design, total economy and contractual economy. Fairly technical and formal parameters, which certainly do not point in a direction of holistic, patient-centered, responsible and hybrid design. This is underscored by the composition of both the contractors groups and the assessment group. The contractors groups are classical and traditional in their composition: architects, engineers, and IT-techies, whereas the assessment group consists of politicians, administrators (hospital and university), medical staff (doctors), architects, and engineers. Specializations and professionals that traditionally take part in such processes, and who are, generally speaking, representatives of an empirical analytical mindset, focusing on systemic control of processes and behaviors. From a postphenomenological and hybrid imaginary perspective we find a disturbing lack of public participation in the process. We also find that emphatic engagement and cross-fertilization is prohibited by the focus on classical specializations. Where are the members of community, because the politicians are not enough on this behalf? Where are the representatives of NGO's in both the contractors groups and the assessment group? Where are experts that can caretake the democratic, social and moral interests of community, like for instance philosophers of technology, techno-anthropologists, sociologists of technology?

Here I would like to highlight one particular aspect in the winning project that elucidates how the ontology of the hospital has not changed radically in a late modern context. The plan of the hospital show how administration and research is placed a part from the various wards and the divider is the huge section of operation theatres and visualizing technologies for analysis: X ray, scanning etc. This means that the vigilant eye and the medical practices of physicians engages with the abstract body of images and the body as machine in operation theatres, where the individual *here-body* of the patient is once again suppressed and dominated on behalf of the disembodied representations ready-at-hand in virtual images or mechanic objects on the operating table. The hospital building mediates this mechanic and abstract organicist perspective of the body (Teyssot 2005) that repeats similar mediations in the pavilion-hospital and in the container-hospital.

The two hospital projects show stunning similarities for what concerns process and design, where is clearly visible an intention to optimize upon economy, facilitate research and healthcare (they are both university hospitals), design for integration of architecture and nature, design for well-being of patients (at least on an intentional level), creating community feeling through the metaphor of the city and iconic architecture. The metaphor of the city and iconic architecture has, as has been written, clear references to early modernism, especially Bauhaus and Le Corbusier, and to the modernist ideals of evidence and rationalism, which might seem in opposition to what contemporary society is in need when it has to confront the challenges of sustainability, society and techno-science. It seems that the design-briefs are focusing on techno-scientific matters and that socio-environmental, socio-ecological and socio-ethical elements are dealt with as add-ons in order to meet emergency of the context.

PERSPECTIVES

Hospitals are mega-machines that accommodate and mediate a variety and complexity of functions, needs, requirements and wishes: cure, care, comfort, healing, health, well-being, life, death, work-place, work-practices, professional performance, other machines practices and performances etc. Hospitals cannot and should not be regarded as mere containers of bodies and machines, but on the contrary as technological organisms that affects our various multistable being together with that organism. This means that the trajectories in multistable hospital design should support possibilities for different and alternative human-technology relations. Ihde pointed out that multistability is not postmodern relativism or scientific objectivism: "Multistability is not the same as neutrality. Within multistability there lie *trajectories,* not just any trajectory, but partially determined trajectories" (Ihde 2002:106). The complexity of these trajectories is obvious due to the amount and diversity of humans, machines, practices, functions, needs, requirements and wishes, which are in themselves complicated and multistable.

In front of this image of complexity, variety and multistability we find that modern man has tried to rationalize and instrumentalize the exploitation and domination of space, and architects of hospitals have to a very high degree contributed to the constitution of the technical and evidence-based paradigm. The performativity of space, technologies, practices and procedures is inscribed in a logic of accountable effectiveness and basically looking at the physical/biological body of the human, which paradoxically is the disembodied *there-body* in Ihde's terminology.

I have been pointing at the twentieth century hospital as a hierarchical technology both for what concerns the vision of the body and for what concerns the institution of the hospital. Despite the fact that recent focus has been placed on the patient with the notion of "patient-centered" care, design etc., then the hierarchy is still governing care procedures, design-processes and political decision-making in hospital design, which the two contemporary cases from a Danish context clearly show. The technological and hierarchical vision of the "patient-in-the-center" is to be compared to how the central-perspective in Renaissance art pushed the center away from the viewer and furthermore kept the viewer and the center in a fixed and static position, or what Herbert Marcuse calls "one-dimensional" perspective (Marcuse 1964). Ihde tells us that it need not necessarily be so and that technology actually has the potential for supporting multiple and multistable "patient-in-the-center" practices and solutions. Furthermore Ihde makes us understand that the "patient-in-the-center" is not a fixed and static entity, but on the contrary in transition, because moving from one condition to another and the trajectories for this movement are and should be multiple.

Ihde points at a possible solution for how to deal with complexity in research and design-processes that might support sustainable and responsible trajectories: "They do (heuristic suggestions) imply that we need to have a deep insight into both technological structure and the history of technologies, which is best based on broad and interdisciplinary knowledge; that a critical take is called for, detracted neither by utopian or dystopian aims; and that multiple variant approaches are likely to be the most promising for contemporary complexities" (Ihde 2002:112).

Hybrid imagination tries to find appropriate solutions of the in between or in the middle-ground (Ihde 2002) and give way for multistable and sustainable solutions and conditions that enhances both the individual and the social body or to rephrase Merleau-Ponty: the aesthetic, ethic and physical body.

REFERENCES

Alcoff, L. M. (2005): "Foucault's Philosophy of Science" in *Continental Philosophy of Science.* Gutting, F. (ed.) Oxford; Blackwell Publishing.

Bachelard, G. (1958/1994): *The Poetics of Space. The classic look at how we experience intimate places.* Boston; Beacon Press.

Bynum, W.F. et al. (2006): *The Western Medical Tradition* 1800–2000. New York; Cambridge Universtity Press.

Dewey, J. (1934/2005): *Art as Experience.* New York; The Penguin Group.

Dirkinck-Holmfeld, K. and Heslet, L. (2007): *Sansernes Hospital.* (The Hospital of Senses) Copenhagen; Arkitektens Forlag.

Foucault, M. (1963/2003): *The Birth of the Clinic. An Archaeology of Medical Perception.* London/New York; Routledge.

Gadamer, H.-G. (1960/1992): *Truth and Method.* 2nd revised edition. New York; The Crossroads Publishing.

Hardy, A. and Tansey, E.M. (2006): "Medical Enterprise and Global Response. 1945–2000" in *The Western Medical Tradition. 1800–2000.* Bynum, W.F., Hardy, A. Jacyna, S. Lawrence, C. and Tansey, E.M. (eds.) Cambridge (UK); Cambridge University Press.

Hughes, T. P. (2004): *Human-Built World. How to Think about Technology and Culture.* Chicago; The University of Chicago Press.

Ihde, D. (1990): *Technology and the Lifeworld.* Bloomington; Indiana University Press.

Ihde, D. (2002): *Bodies in Technology*. Minneapolis; University of Minnesota Press.

Jamison, A., Christensen, S. H. and Botin, L. (2011): *A Hybrid Imagination. Science and Technology in Cultural Perspective*. London; Claypool & Morgan Publishers.

Jonas, H. (1979/1984): *The Imperative of Responsibility. In Search for an Ethics for the Technological Age.* Chicago; The Chicago University Press.

Leder, D. (1992): "A Tale of Two Bodies: The Cartesian Corpse and the Lived Body" in *The Body in Medical Thought and Practice.* Leder, D. (ed.) Dordrect/Bosotn/London; Kluwer Academic Publishers.

Marcuse, H. (1964/2002): *One-Dimensional Man. Studies in the ideology of advanced industrial society.* London and New Your; Routledge.

Merleau-Ponty, M. (1945/1994): *Phenomenology of Perception* New York/London; Routledge.

Teyssot, G. (2005): "Hybrid Architecture: An Environment for the Prosthetic Body" in *Convergence: The International Journal of Research into New Media Technologies vol. 11(4)*:72–84.

Verbeek; P.-P. (2005): *What Things Do. Philosophical Reflections on Technology, Agency and Design.* University Park (Penn.); The Pennsylvania State University Press.

Verbeek, P.-P. (2011): *Moralizing Technology. Understanding and Designing the Morality of Things.* Chicago; The Chicago University Press.

Williams, R. (1989/2007): *Politics of Modernism. Against the New Conformists.* London; Verso.

Chapter 7

Post-Telescope-Postphenomenology? . . . and a Little Locomotive History

Michael Funk

Question:

What is the future of philosophies of technologies? What could be the next steps of postphenomenological research?

Answer:

One more post: *post*-telescope-postphenomenology!

IN SEARCH FOR A *POST* . . .

Philosophy emphasizes human subjectivity related to concrete theoretical problems and practical challenges. One of those challenges, theoretically as well as practically, belongs to accurate and critical investigations of technologies, technics, and technique (Mitcham 1992). Like in any other case of human life, also in philosophies of technologies an understanding of the meaning of "subjectivity" is required. At this point we are enabled to grasp the first *post*, because it draws attention to the replacement of a so-called "transcendental subject." Phenomenology in the classical sense remains *pre-post* as long as its philosophical roots can be found in the idea of a disembodied pure subject in opposition to a pure object. Clemens Brentano and the early Edmund Husserl belong to those *pre-post* movements. Both became something like godfathers of classical phenomenology. In *post*phenomenology the transcendental subject is replaced by an embodied subject of cultural practice. Maurice Merleau-Ponty and Martin Heidegger are two important idea generators (so to say "godfathers of a second generation"). And even

Edmund Husserl himself belongs to this line of philosophical thought as he turned to a more pragmatic lifeworld-thesis in his late book *Die Krisis der europäischen Wissenschaften und die transzendentale Phänomenologie* (a fragment, based on lectures by Husserl in the 1930th). Metaphorically we may say that Husserl is a perfect dualistic thinker: he remains godfather for classical phenomenology (*pre-post*) even when he became later a second-generation-godfather (*post*) together with Merleau-Ponty and Heidegger. In proximate developments, basic postphenomenological concepts have been elaborated by Don Ihde (1979, 1990, 1991, 1993, 1998, 2015). As far as I see, most important insights are:

1. Technique, technics and technology is *technical praxis*
2. Technical praxis is *embodied praxis*
3. Embodied praxis is *cultural praxis* into a concrete lifeworld
4. Embodied cultural praxis is related to *embodied processes and activities of knowing*
5. Embodied processes and activities of knowing constitute *technical mediations*—that is the way in which instruments shape human-world-relations and sensory perceptions (e.g., the ways in which telescopes and microscopes reveal invisible macro- and micro-worlds)
6. Embodied processes and activities of knowing also cause *multistability*—that is the way we use specific tools and processes in different ways (e.g., horses and the tool "breeding" can be used for agriculture, war, cooking, as status symbol, for sports and so on; a more common example is the knife: it can be used for butter and bred but also for killing people)
7. Embodied processes and activities of knowing enable *technoscience*—that is the way technological practices and scientific researches are inherently intertwined
8. Technoscience emphasizes no pure realism but *instrumental realism*
9. Instrumental realism is not related to subject-object-dualism or a Cartesian ontological dualism of res extensa (the physical body) and res cogitans (the pure cognition)—as already mentioned: the disembodied transcendental subject is replaced by an embodied subject of concrete bodily and cultural *technical praxis* (here we close the trajectory from 1 to 9)—reality is practical sensory-bodily constituted and belongs to the usage of tools and instruments
10. All that is characteristic for postphenomenology and at the same time for *expanded hermeneutics*—both concepts go (bodily spoken) hand in hand but are (geometrically spoken) not congruent to each other. Hermeneutical postphenomenology is not the opposite of natural sciences. This understanding belongs the *pre-expanded* meaning of linguistic and text-oriented hermeneutics in the tradition of Schleiermacher and others. Hermeneutical postphenomenology is philosophical

framework of bodily and sensory interpretations and actions. Don Ihde's approach is not only bridging the gap between natural sciences (Naturwissenschaften) and social sciences (Geisteswissenschaften)—it is also compatible to many forms of cultural practices such as handcrafts or arts. Postphenomenological investigations also focus on musics, the relation between string instruments and archery or recording technologies and whale sounds. There are differences and similarities between scientific instruments and musical instruments as it has been emphasized by Don Ihde as well (Ihde 2007, 2010, 2013b).

IN SEARCH FOR ANOTHER POST

What could be a second *post* related to "classical" postphenomenology—a *post*-postphenomenology? *Pre-post* would mean to fall back behind the second godfathers (Merleau-Ponty, Heidegger and Husserl II), but *post-post* enables new perspectives and progressive approaches in the twenty-first century. I do not want to summarize all aspects of the wide field of potential *post-post*-developments. May be I am not able to overview it completely. And honestly spoken: we also should not even try to overdo the *post-post-post-post* . . . chain. Hmmm . . . Or may be we should exactly try it—as an intellectual experiment! Anyway, it is my emphasis to focus one concrete piece in the wide field of possible *post-posts*: the locus classicus and maybe also the "holy grail of postphenomenology": Galileo and his telescope! Examples and conceptual analyses of mediating technologies or material hermeneutics and technological embedding of scientific research are often illustrated by focusing Galileo Galilei—not primarily as mathematician (this would mean *pre-post*) but as concrete human and embodied being with a concrete technical tool, the telescope, that shaped his praxis and perceptions into a concrete lifeworld.

I do not want to disgrace this "holy grale of postphenomenology" as I also belief the concept of technological mediation is true. But it is a truth that is worth to be scrutinized to enable new perspectives. Symbolic speaking: why not also focus on van Leeuwenhoek and his microscope? Don Ihde did also that (currently in Ihde 2013a), but we should not forget that Galileo was mainly a physician and not a biologist. Is there a methodological and conceptual difference between technical mediation in physics and biology?

I think that main approaches in the philosophy of sciences (whether or not they are *pre-post* or *post*) tend to be physic-centric. Physics is seen as the ideal of modern sciences. This physic-centrism started in early modernity with Galileo, Descartes, and others, and it seems to be characteristic on a conceptual level even for postphenomenology. Now it is at the time for a non-physic-centric postphenomenology and that means it is time for a *post-telescope*-postphenomenology! Therefore I want to perform a well proofed praxis: a case study. Focus is on twentieth-century railway and locomotive history, but it may teach us something for *post*-telescope-postphenomenology.

RAILWAY UPSIDE DOWN / LOCOMOTIVE FRONTSIDE BACK

This little case study is related to an observation, which I made in the *California State Railroad Museum* in Sacramento. When we think about a steam locomotive, we are used to imagine something like that:

Figure 7.1 A classical nineteenth-century steam locomotive. *Source*: Photograph: Michael Funk (private archive)

In classical design, the engine is situated front side and the wagon with fuel—that has been coal for a long time—comes behind the engine. Between engine and fuel lies the cockpit. There is at least one very pragmatic reason for this structure: locomotive engineers need to be able to shovel the coal by hand into the engine. So far the cliché of nineteenth-century and early twentieth-century steam locomotives . . . Going on walking through the museum later on I made non-cliché observations because of a steam locomotive that looks like this:

Figure 7.2 The cab-forward Southern Pacific Railroad Locomotive No. 4294. *Source*: Photograph: Michael Funk (private archive)

As one can see, the cockpit's situation is front sided, the engine lies immediately behind the cockpit and behind the engine is the fuel. I became curious because of the peculiar design and one guy, dressed like a locomotive engineer (you can see him in the white dress standing into the cockpit), started explaining the story of *Southern Pacific Railroad Locomotive No. 4294*. It was used between 1944 and 1956, and this one in the museum is the last surviving so called "cab-forward" steam locomotive. Locomotives like that have been primarily used in mountain-tracks such as the Sierra Nevada. Railway tracks in the mountains are especially in the winter a genuine challenge for steam locomotives—not only because of the ups and downs. Moreover the snow is a challenge; a challenge that has caused the massive application of wooden snow sheds. Imagine a railway track in the mountains that is a combination of tunnels and roofs for many, many, many miles! What do you see? Exactly: Smoke in the locomotive engineers faces! As the guy in the museum explained, that was the environmental caused reason for cab-forward design.

That observation illustrates how technological developments are also caused by "natural" environments, even if we focus modern technologies such as steam locomotives. But we can realize also at least one more thing: the impact of technological developmental paths (Irrgang 2001, 2002a, 2002b,

2007). Why was the cab-forward design applied since the 1940th and not in the nineteenth century (or even earlier)? Of course, the guy in the museum had an answer for that. From a technological point of view it was not possible to carry the coal from the rear wagon over or under the engine into a front sided cockpit. Coal consists of bricks—it is not fluid. The technological solution was related to general developments of new engines on basis of fueled oil. In the 1940th fueled-oil-engines have been available and so it was easy to pump to fluid fuel from the rear wagon by the engine into the cab-forward cockpit. Instead of coal, the locomotive-engineers inserted fueled oil from the cockpit into the engine. Insofar the "user-interface" of the *4294* remained that one of a classical steam-locomotive. What do we learn? A non-locomotive developmental path of fueled-oil-engines caused a solution for an old environmental-related problem in the design of steam locomotives. Technological developmental paths are not single isolated processes but shaped by each other and embedded into a more complex framework of cultural knowing and technical applications.

There are many more examples for that. I just want to mention something I have observed in the *Danmarks Jernbanemuseum*, the Danish railway museum in Odense. Of course, there are many locomotives, wagons and models of railway history in this museum. That is what the visitor of a railway museum anticipates to see. What surprised me, I did not expect to see something like that in a railway museum, was a room full of ship-models that illustrated the history of Danish shipping. What is the relation between ships and locomotives in Denmark? The answer is easy if we focus the map of Denmark: it is a country with many islands and the capital, Copenhagen, is located on the island *Sjælland* at the very east of the country. As long as there have been no bridges or tunnels ferries have been used to carry the locomotives and wagons from one island to another. The development of Danish railway goes hand in hand with the development of ferry boats—something that one would not expect from a US-American point of view (because railway was made for long distance and connections inside a continent) and from my German perspective this was surprising as well. From a philosophical point of view this example illustrates the same insights: the impacts and interrelations of technological developmental paths (locomotives, ferry boats and later high-tech bridges or tunnels); and the influence of "natural" environment at the same time (mountains and snow sheds in California; islands and ferry boats in Denmark). Coming back to the *California State Railroad Museum*. One more observation can be added: sensory mediations.

RAILWAY HISTORIES AND VISUAL HISTORIES

As emphasized in the Californian museum, the history of American railway is also a history of the transcontinental connection between east and west coast—mainly in the 1860th. That history is not only about locomotives engines or railway tracks. It is also a story focusing another technological developmental path: photography. First photo cameras have been heavy and bulky. All in all the complete equipment was about 100 pounds or 45 kilo. But those cameras enabled new perceptions and perspectives on railway developments as they bore new photorealistic mediations and visualizations of what has happened. In the museum this was pointed within the slogan "Two Views of History" (the photograph and the painter). A main event was 10 May, 1869: the meeting of Union Pacific (starting from Omaha) and Central Pacific (starting from Sacramento) tracks in a place today called *Golden Spike National Historic Site* (Utah, close to Salt Lake City). Final connection of the transcontinental railway has been staged for the newspapers by photographer A. J. Russell. The development of photography shaped also the image and visualization of railway developments—its social and historical perceptions:

Figure 7.3 Yes, this is also postmodern pluriculture: a German guy in a Californian railway museum taking a picture of the reprint of a nineteenth-century-photography by A.J. Russell. . . . *Source*: Photograph: Michael Funk (private archive)

On the other hand, also traditional historical painting has been applied as well. In 1877, Thomas Hill was accomplished to paint the same event and thereby he also followed the instructions of the client, changed the setting, added and removed persons from the scene. Insofar, there is something like a visual multistability of this historical date. One perceptual interpretation is provided by classical painting and another is enabled by technologies of photography. And there is even more sensory multistability: the pictures can be interpreted in terms of historical research or museum education; but they can also be interpreted in terms of a post-telescope-postphenomenology (that is what I try to do now).

Of course, nineteenth- and mid-twentieth-century railway examples may sound old fashioned. Today we live in the age of webcams, digital photography, smartphones and the internet. We live in sensory lifeworlds with many multistable perceptual interpretations of peculiar events or processes—social as well as historical. One conceptual term used by Don Ihde is "pluriculture" (Ihde 1993). It describes the very different cultural pieces and traditions that are mixed up within new image technologies and also bear potentials for new cultural synergies. But let us also try to reinterpret this insight in terms of environmental related sensory and cultural self-organizations. Pluriculture is not only a word to describe cultural developments, it also describes organic-biological potentials as we are bodily (*leibliche*) beings that are also shaped by organic-cognitive competences and skills in handling, applying and using those sensory multistabilities.

SOME CONCLUSIONS

Now, what can we learn form that locomotive and railway histories about a post-telescope-postphenomenology? First, I must admit that railway also tends to be a physic-centric example and a symbol for industrialized modern technologies. One may argue that examples of industrialized technologies are not sufficient for illustrating today's more hypermodern information technologies, autonomous technologies or biotechnologies. That is true. But I also think we can use such examples for establishing at least some new interpretations of technological praxis and developments. Hypermodernity does not mean that today's technologies are completely new; moreover they can be understood as continuation of modern developments (Borgmann 1984, Irrgang 2008).

One conceptual conclusion is that we are able to relate sensory multistability to technically mediated human-environment interactions in terms of epigenetic developments. New genetics provide new perspectives on the organic structures of human embodiments. And with respect to paleoanthropology it becomes obvious that human evolution is a peculiar combination of natural and cultural evolution (Irrgang 2009, Schrenk 2008). The cab-forward locomotive is a simple example for that, because it illustrates how even the development and use of industrialized technologies such as classical railway can be understood in terms of a feedback-loop between cultural and natural developments.

Here we are able to find a strong link to postphenomenology and expanded hermeneutics: technology is culturally embedded technical praxis and interpretation. Human embodiment, the human body (*Leib*), from a genetic and epigenetic point of view is itself inherently multistable. There is a peculiar feedback-loop between the observer's perspective position and the observed structure, which can be described in conceptual terms of self-organization or autopoiesis (Maturana & Varela 1992, Irrgang 2015). Instrumental praxis, that started with first hand axes and easy handcraft-tools more than 1 million years ago, and human biological organic embodiment are interrelated—even today. Cultural embedding means a challenge for learning new skills—from a cultural point of view, new technologies are not behind a non-bridgeable gap. Here is the place for a *post*-telescope-postphenomenology. It remains philosophical reflection as its focus is on human cultural practice/praxis. This is one side of the coin—cultural evolution. The other side is provided by natural sciences such as epigenetics, paleoanthropology or the new biologies—natural evolution. A methodological carrier of both sides is transclassical hermeneutics—the competence of circulatory interpretation and interrelating both poles (Irrgang 2009). A post-telescope-postphenomenology without any natural sciences remains empty, such as natural sciences without any post-telescope-postphenomenology remain blind (to play a little Kantian language-game). *Hermeneutical post*-telescope-postphenomenology means both.

If the slogan "hermeneutical post-telescope-postphenomenology" is too bulky, one can easily call it "philosophy of technologies." Following this intellectual "trajectory" (to use a metaphor often applied by Ihde) the toolbox of postphenomenology and expanded hermeneutics can be enhanced and further developed related to current technological developments. Whether or not we like to label such philosophical reflections "post-telescope-postphenomenology," what matters are the critical and philosophical skills behind those words.

REFERENCES

Borgmann, Albert 1984: Technology and the Character of Contemporary Life. A Philosophical Inquiry. Chicago & London.

Ihde, Don 1979: Technics and Praxis. A Philosophy of Technology. Dordrecht a.o.

Ihde, Don 1990: Technology and the Lifeworld. From Garden to Earth. Bloomington a.o.

Ihde, Don 1991: Instrumental Realism. The Interface between Philosophy of Science and Philosophy of Technology. Bloomington a.o.

Ihde, Don 1993: Postphenomenology. Essays in the Postmodern Context. Evanston, Ill.

Ihde, Don 1998: Expanding Hermeneutics. Visualism in Science. Evanston, Ill.

Ihde, Don 2007: Listening and Voice. Phenomenologies of Sound. second edition. New York.

Ihde, Don 2010: "Technologies-Musics-Embodiments." In: Ihde, Don 2010: Embodied Technics. Copenhagen, pp. 17–36.

Ihde, Don 2013a: "Ricoeur and material hermeneutics." Speech in honor to the 100dth anniversary of Paul Ricoeur's birth (unpublished).

Ihde, Don 2013b: "Embodiment: Technologies and Musics." In: De Preester, Helena (ed.) 2013: Moving Imagination. Explorations of Gesture and Inner Movement. Amsterdam & Philadelphia, pp. 101–112.

Ihde, Don 2015: "Euro-American Philosophy of Technology Today." In: Funk, Michael (ed.) 2015: "Transdisziplinär" "Interkulturell." Technikphilosophie nach der akademischen Kleinstaaterei. Würzburg, pp. 53–70.

Irrgang, Bernhard 2001: Technische Kultur. Instrumentelles Verstehen und technisches Handeln. Philosophie der Technik Bd. 1. Paderborn.

Irrgang, Bernhard 2002a: Technische Praxis. Gestaltungsperspektiven technischer Entwicklung. Philosophie der Technik Bd. 2. Paderborn.

Irrgang, Bernhard 2002b: Technischer Fortschritt. Legitimitätsprobleme innovativer Technik. Philosophie der Technik Bd. 3. Paderborn.

Irrgang, Bernhard 2007: Hermeneutische Ethik. Pragmatisch-ethische Orientierung in technologischen Gesellschaften. Darmstadt.

Irrgang, Bernhard 2008: Philosophie der Technik. Darmstadt.

Irrgang, Bernhard 2009: Der Leib des Menschen. Grundriss einer phänomenologisch-hermeneutischen Anthropologie. Stuttgart.

Irrgang, Bernhard 2015: "Skeptisch-kritische Epistemologie, kontextbezogene Selbstorganisation des Verstehens und positional-perspektivische Metahermeneutik." In: Funk, Michael (ed.) 2015: "Transdisziplinär" "Interkulturell." Technikphilosophie nach der akademischen Kleinstaaterei. Würzburg, pp. 463–531.

Maturana, Humberto R. & Franciso J. Varela 1992: The Tree of Knowledge. The Biological Roots of Understanding. Boston.

Mitcham, Carl 1994: Thinking Through Technology. The Path between Engineering and Philosophy. Chicago.

Schrenk, Friedemann 2008: Die Frühzeit des Menschen. Der Weg zum Homo sapiens. 5. A. München.

Chapter 8

The Dubstep Mashup

Stacey O. Irwin

Electronic dance music has, in some form, been around for close to fifty years. The current iteration of the genre, often called EDM, is produced primarily to be played in nightclubs or concert venues and warehouses, and is largely created for DJs who remix electronic music in a variety of ways that evoke a live performance feel. This music, along with its performance component, has been articulated in a wide array of research (VanVeen and Attias 2011, Pfadenhauer 2009, Dekker 2003). One of the more recent movements, or subgenres of EDM, is called dubstep. This community is enthusiastic and passionate and their beats and hooks are infectious but the history of this genre is contentious and what constitutes "dubstep" is under continual scrutiny on fan sites and popular media. Many agree, however, that dubstep is the sounds of a generation. Strands of DJ performance, a new language for digital tools and music composition, and DJ and fan collaboration will be considered in this philosophical mashup. The notion of mashup in the title is used in the widest sense, to contextualize the idea that this reflection pulls from a wide range of sources like philosophical texts, magazines, websites, personal DJ experience, that becomes transformative of the original text to provide a new collaborate whole in the understanding of dubstep. Notes Don Ihde, "A phenomenological analysis or description, as it is technically called is more than mere analysis. It is a probing for what is genuinely discoverable and potentially there, but not often seen. Phenomenology is the door to possibility, a possible that can be experienced and verified through the procedures that are, in fact, the stuff of experimental phenomenology (2012:13). This reflection hopes to discover what is potentially there, buy not often seen or heard, from the "dubstep nation."

"The Thirty Greatest Dubstep Songs of All Time," Spin. July 24th, 2012. http://www.spin.com/articles/making-brand-40-greatest-band-names-all-time/.

WHY STUDY DUBSTEP?

My turning to the phenomenon is partly personal curiosity. I hear dubstep around me, in clubs, in movie and videogame soundtracks, and in commercials. Music magazines and websites talk about the "Godfather" of dubstep and the "early adopters" of the genre, while mainstream music has only begun to embrace the genre. Dubstep is not regularly topping the music "hit" charts but has been a success at clubs and on various sharing websites for a decade, and sites like *beatport* rate the top mixes of the genre. The genre is neither underground nor established. It is making its own mark and own independent market for music while avoiding mainstream media in favor of the becoming the kingpin of the club scene.

My DJ friends are performing dubstep mixes at clubs and parties and I have been considering how this kind of music is different from previous forms of electronic music. And I must confess, I have always had an interest in sound. I've been a DJ at several radio stations and taught digital audio production classes at my University. My work in digital media allows me to "play with" a variety of digital audio tools and I had the opportunity to learn to use a MIDI in a graduate school class to create a sound bed for some visual work I created. I enjoyed the rave scene in the 1990s but stopped clubbing to get my doctorate and have three children. And truth be told, I married the sound guy in my brother's band. So I turn to this phenomenon with sincere interest, curiosity, and a bit of understanding. Ihde (1986) shares that "[A]ll fields of possibilities display some type of topography, which becomes apparent only when the field itself is opened to the essential insight of the rigorously descriptive phenomenological look" (136). This reflective essay only scratches the surface of a complete study, but explores the phenomenological attitude as an opening for further study. And lastly, the study of dubstep is an opportunity to "take up" Don Ihde's texts that I have learned much from, to explore this digital music phenomenon.

To be sure, any study of material technological artifacts *The Ihde Way* comes with some caveats. First, notes Ihde (1993), all technologies are non-neutral. "They are transformational in that they change the quality, field, and possibility range of human experience, thus they are non neutral" (33). Second, "technologies must be understood *phenomenologically* i.e., as belonging in different ways to our experience and use of technologies, as a human-technology relation, rather than abstractly conceiving of them as mere objects" (34). And third, the technological transfer is multidimensional in that it will "involve basic cultural and existential interchange" (34). One additional layer deserves mentioning: an understanding of the intentionality or "trajectory" of technologically embedded experiences. The myriad of technology used in the creation of dubstep mediates the relationship between DJ's and their fans and listeners, and plays an active role in its creation and context.

> When technologies mediate the intentional relation between humans and world, this always means from a postphenomenological perspective that they codetermine how subjectivity and objectivity are constituted. Their intentionalities, in Ihde's sense, consist of the fact that they co-shape the contact between human beings and their world: they determine how human beings can be present in the worlds, and the world to them. (VerBeek 2005, 6)

The mediation of the subjectivity and the objectivity is a key element in this reflection. How DJs and their technology co-determine and co-shape dubstep is an important consideration in the DJ (human) turntables, computer, sharing websites (technology) fans, audience, sharing websites (world) relation.

So with these foundational elements in line, I've come to a place where I am wondering about the human-technology-world relation in dubstep as part of the cultural milieu, and not just as a social experience of listening, but as an experience of technological co-shaping. This reflection aims to think about several of the many variations of the dubstep genre, to better understand this very popular current technologized music trend while "respecting and yet seeing a deeper multistable structure to these possibilities"(Ihde 1993, 71).

The dubstep environment is one of listening, mixing, remixing, performing, live mixing, listening, moving, remixing the live mix, dancing and experiencing. So I wonder, how do dubstep creators imagine their music? Has dubstep become a new cultural context for the technology used for this genre? Has one of these "post" genres called brostep, a more Americanized version, become the technology transfer based on a new set of material artifacts available to DJs, the largest group of dubstep music makers? And what of the newer iterations called brostep, post-dubstep and purple dubstep? For Ihde, (1993) technology is a cultural instrument only limited by the cultural imaginings of the individual. Indeed, the material music culture of dubstep, through human-technology-world interaction and use, has imagined and created new cultural dimensions.

DIGITAL BEGINNINGS

In the mid to late 1990s a group of DJs started experimenting with new computer-based sounds and inputs and outputs in garages and homemade studios in South London. The digital sounds eventually trickled into warehouse dances and clubs to the point where other digital music creators, though less sophisticated in the ways of building these digital mixes, started experimenting too. Explains a 2012 *Spin* magazine article:

> The sullen spawn of U.K. garage and British sound-system culture, dubstep first took root a decade ago in London club nights like FWD (and a couple years later, DMZ), where chest-hollowing bass lines met Spartan breakbeats under

> the invocation: 'Meditate on bass weight.' Maybe it was all that low-end pressure that made it explode so quickly: By the late 2000s, dubstep had splintered into numerous factions, from brostep to wonky to the evocative "purple," even rumbling stadium audiences with earthquake drops. (7/24/12)

After dubstep found a homeland in the UK, it spread throughout the world and created subgenres that shadow but do not reify the original goal, while adding new and different dimensions to the sounds. The combination of affordable equipment components of varying levels of complexity, website platforms to post dubstep for sharing, a ready crowd to critique the music on sites like SoundCloud, beatport and subnav and sharing on social media and Youtube once a DJ's set was ready for the big reveal, made dubstep a popular pastime for both creators and fans.

Like a band, Dubstep is created from a set of songs, beats, and previously recorded "mixes" built from a loose playlist based on the interest, requests, and swell of the crowd as an organic feedback mechanism. But crowd swells at live concerts have been around for a long time. Film clips of the Beetles, Rolling Stones, Elvis and U2 bear witness to the power of crowd enthusiasm and interaction. What makes dubstep different is the continued doubling over and remixing of the raw material into a derivative work. This synchronized segue, an individualized mix, is often connected with the 1990s rave dance trend, and is the cousin or child, depending on who you ask, of disco, house, techno, hip hop and Jamaican DJ cues. Popular dubstep artist Skrillex, who had collaborated on several projects with artists who largely work in other genres, explains it this way:

> It just lets so many people in and there's nothing about it that seems shoved down anybody's throats. You can connect with it culturally because it brings so many different types of people together, from ravers to hip-hop people to whatever people like to dance. (Herrera, 2011)

The combination of already created music and new sounds, remixed and recreated in different layers live by a DJ, moves the production and mastering phase into perpetual loops as the performance of the materials becomes part of the digital mix. That digital mix is again tweaked, uploaded, shared, potentially downloaded by others to use in a mix, and then possibly played live again for another loop of digital interaction.

At its essence, dubstep is a music mix combining the use of sequencing platform software programs like Logic Studio or Dub Turbo, with an Auto-Tune plug in, and speakers with large sub-woofers to hear the bass and drums, a beat box with a 140 beat per minute timing, and the chorus of a song or two as a hook to keep people familiar with the song as they react to it within the club environment or venue.

Part of the allure is wobble bass or web, used frequently in dubstep, as an extended sound that takes on a rhythmic tone. A low pass filter with a low frequency oscillator assigned to the cut-off proliferates the distinct wobble of the bass, which combines with the drum. The sound is facilitated by the use of large speakers technologically able to transform the low sound into amplification for the audience. Early dubstep especially, is rarely played through earbuds, because much of the low wobble of the sub bass is lost because the tiny speakers are technologically unable to transfer the low tones of dubstep that can be felt by listeners in a very bodily way. At its best, dubstep is a kind of music created to be heard in environments that have the speakers that are able to capture and amplify that low sub-bass range.

MIX, MASTERED

DJ, Technology, Sounds and the audience co-create dubstep. The variety of technologies is used in the digital creation, but the final mix is output on social media sites and uploaded onto dubstep sites for regeneration into another DJ's mix. These mixes are played live at clubs and in the studio. The mixes become fluid interweavings of what is created, what can be used from others, and what can be manipulated within one's own dubstep workstation. Dubstep uses many of the same digital software tools and vocabulary as EDM, so the proliferation of what Ihde calls "*familiarity within a known praxical gestalt*" (35) occurs within the dubstep movement to create subgenres. The consistently recognizable and understandable interface of the variety of music mixing tools available like propellerhead, Deckadance, Live and Push, digital audio workstations like FL Studio, and many free downloadable sounds, paired with easy creation of MP3s uploaded to sharing sites, fosters the genre and EDM in general. Synthesizers, drum machines or beat boxes, sequences, and synthesizers, along with vocabulary like sampling, scratching and tracking, drops, and hooks, have become universal to DJs, fans, electronic music brands, and the musicologists, historians, and cultural anthropologists who study digital music.

The compound hearing experience of dubstep fragments, feedback loops and electronic generated sounds, pulled together in space and time, constructed on the fly, from components mixed over time and new elements, means song turns into set and turns into a twenty minute "mix" or entire "set" of music occurs before a "break." One thing is for sure, a population of opinionated dubstep loving people is inhibiting the planet and have probably already begun to disagree with the ways in which I have explained dubstep or categorized its place in EDM history. And that is par for the course for the genre and its lively fan base.

THEORY, CONSIDERED

Variational theory, which uncovers the variety of multistabilities of material technological artifacts, is one of the key concepts of this reflection. Ihde employed the concept of multistability as a way to explain technological mediation. This insight paved the way for postphenomenological research as an approach to understand the transformation of the perception of an object. The idea is to uncover variations in the mediating role of dubstep for the audience and the DJ to reveal the multiple ways that the music technology facilitates, fosters and creates dubstep, in sync with the embodied DJ and audience, to create a "new" music mix, that is posted, downloaded, and again doubled over on itself as a new performance experience. The mixes that are hits, and uploaded, voted on, and downloaded and used again, in a way that erases much of the ownership of anything original, for the newness of the most recent mix.

The work emphasizes essential structures or essences (Ihde 2009) to determine what is variant and altered and what is invariant or unaltered. It is through beginning insights of a technology's use, that we can have, notes Ihde, "insight into the very structure of multistability, an insight that then can guide our subsequent awareness, such that we might well expect both more possibilities and . . . multiplicity of profiles" (Ihde 2006, 201). Thinking this through can help develop a way of constituting ourselves in this mediated world, to reveal a deeper and more rigorous analysis of the illusion of the digitally altered body engaged in the wub, wub, wub, wub of bass, with drops and pops of the hook, loops, amidst the strain of a pop song.

Postphenomenological variation and inversion can be uncovered by thinking about the microperception and macroperception of the music genre, with its focus on bass, griminess, dirty sounds, and hook. A microperceptual lens highlights the sensory experience of feeling the beat of the bass, recognizing a familiar hook, moving in sync with the crowd, and having fun in a culturally attuned ritual of dance and movement. The gaze is intent on taking the experience in, viewing the lights, which often mirror the beat and audience movement. Macroperception, conversely, discovers that this specific perception, this sound, this view, this look, this body, can be completely appropriate and attainable. Contextualized, there is a cultural shift of accepted experience.

What is revealed in the essences or variational elements of *performance, new language,* and *collaboration* through the experiences of those who dubstep?

Source materials for this reflection have been gathered from magazines, Internet interviews, websites and a transcribed interview to consider the dubstep experience. Experiences of EDM and dubstep performer and producer, California native Skrillex, Popular Bristol, England duo Dodge and

Fuski, Unsigned New York, New York Club DJ TIMarbury and South Central Pennsylvania local spinner DJ Merc provide a context for the dubstep experience. Articles in *Spin*, *Rolling Stone Magazine* and Dubstep 101.com also provide specifics about how a DJ gets started and the components and vocabulary for the genre. Together, multiple ways of relating and communicating in and through dubstep can be considered.

DUBSTEP AS . . .

Language

Like any other cultural movement, dubstep has spawned a whole new vocabulary of terms. Reading fan comments on a website like SoundCloud is one way to explore the variety of vocabulary associated with dubstep. Many different blog posts also reference the music side of the genre (glitched sequences, mini mix, samples, rave synth, trance, trap slaps, beats, pops, bounce, riffs, wub, jams, tempo, dirty, grungy, muddy) as well as the technical side (beat pad, track, leads, samplers, midi controller, 140 bpm, kick sample, punch, patch, sub, reverb and drop). DJ Merc explains it this way:

> There is a really constantly loud predominate bass line, with pretty synthetic synthesizer noises, auto-tune beats throughout the song and obviously it goes to a tune, but there's no lyrics, its all noise that make up a song and they can actually beat mix on the spot. . . throw in the hook.[1]

DJ TIMarbury shares about this equipment:

> For my programs, I use FruityLoops Studio 9 XXL Producer Edition, which a lot of people might say is noob or amateur but, believe it or not, you can use it and get very high quality results. I use various VSTi plugins that I bought and found all over the internet. Major ones are Native Instruments Massive, Syleth, Harmless, Predator, Monomate, Peach, Triforce, z3ta+, ReFX Nexus for synths.

His use of dubstep vocabulary like noob (new DJ) and the variety of different software components illustrates the cultural side of the making.

Performance

Dubstep is at its heart, a music performance. The DJs are digitally performing the dance of audio editing. They mix the beats and bytes and bits to the rhythm and cadence of interwoven sounds. They perform their understanding of the technology by negotiating the place of the software by pulling in different sound strands, laying them down on a digital track interface, and then mixing them and compressing them to a MP3 or other sound codec to be reduced in file space and played on a website for comment and critique and listening or performed "live" for crowd. In a Dubstep 101 interview, Dodge and Fuski explain the elements they bring into their performance of dubstep.

> Loads of stuffit was quite an experimental EP; Violins, a circuit bent answering machine, an old vintage drum kit we had lying around in our old studio and Cubase 6 to name a few! As usual there are quite a lot of use of Native Instruments products used.

The DJ producers knit old and new, instruments like violins with dubstep machines like the Cubease 6 and Native Instruments. This gathering of sound acts like an open playing field of endless perceptual options of "experiments" using things "lying around." This is "*starting in the midst of* that which is already here, the already constituted. That is a beginning from the center" that is dubstep (Ihde 2007 48). A dubstep DJ listens "to the *voiced* character of the sounds of the World" (47). DJ TIMarbury appreciates that platform and the feedback.

> I can upload any of my tracks I want to play, play them to a live audience, get responses and even promote where people can find more from inside of my own room. Social media changed the world in so many ways, not just the music industry.

For him, the idea of online feedback from inside his room is a valuable and social opportunity. His sense that the online community is "live" suggests different notions of embodiment and interaction. The social element of EDM and other digitally created music has philosophically been questioned and it was considered long before dubstep became a genre. Noted musician John Cage, in his essay The Future of Music:

> Technology has brought about the blurring of the distinctions between composers, performers, and listeners. Just as anyone feels himself capable of taking a photograph by means of a camera, so now and increasingly so in the future anyone, using recording and/or electronic means, feels and will increasingly feel himself capable of making a piece of music, combining in his one person the formerly distinct activities of composer, performer and listener. (John Cage, The Future of Music, 1974)

In the essay, Cage explains his concerns that the blending of the roles will take the social element form the experience, saying that, "to combine in one person these several activities is, in effect, to remove from music its social nature" (2009, 120). But this has not happened in dubstep. The sharing, the listening, and the relooping of the mixes and reloading and resharing, based on audience interest, feeds off the social environment, either live or through digital sharing.

Collaboration

Collaboration is the name of the game in dubstep. There is collaboration in the making, the sharing and the performing. Dodge and Fuski explain:

> It's a collaborative effort really, around 5 people randomly including 2 visiting members of the Danish aristocracy were involved in writing it and was written partially at our farm/studio complex in rural wales and partially at our Bristol studio.

And not the kind where you sign a contract to work together for a specific period of time, but the kind of post Napster social media sharing that floats beats and mixes out through websites to be downloaded, integrated into live events, shows and clubs, and remixed and re-uploaded for listening and critique. Some websites work on a paid subscription basis, where they keep track of uploads and downloads and DJ's get points for their popularity. In other cases, listeners typed comments on the track in real time, noting where the "sick beat" fell.

This collaboration within a community was essential for TIMarbury, who was inspired that even "teenagers" and "normal people" could share work. He appreciated that the sharing site Newgrounds was available to him. Newgrounds section titled "Newgrounds Wiki: About Newgrounds" explains that "Feedback is the fuel that keeps us going."

> Being on Newgrounds, looking through all the music, I said to myself, 'Why can't I start to make music like this?' So I started to talk to the artists on Newgrounds. They were all normal people, some were even teenagers, and I did research online on what programs I could use and I learned a lot really quickly with their help. I started to post music on Newgrounds is 2009 and still do to this day.

TIMarbury was inspired by other collaborators and also from the feedback he receives from his fan base.

> At times I even thought of stopping because I wasn't really moving anywhere, in a slump where the music wouldn't come out of me and what's the chance of some kid who makes music on his computer to become popular, but I get these awesome reviews on Newgrounds from fans and I just can't stop if people out

> there enjoy what I create. I also feel like if you listen to my songs, they are unique. I really can't put my finger on a dubstep or electro producer who I sound like, which I feel is great. I still have a lot of work to do to become better though, I never deny that.

He is clearly a testament to Newground's notion that feedback fuels the creator. He balances his personal doubts against the collective interest, and "can't stop" if people are enjoying his work and unique sound.

DJ Merc has paid subscriptions to several different sharing sites. He explains:

> There are a couple of DJ websites, one is particular is called your remix.com, it is a sharing website. I can download as many beats as I want but I have to upload ten original mixes myself. And it was nothing but this DJ website that everyone belonged to and we shared our mixes and downloaded other peoples but they could download from us and at the end of the month they would let you know how many downloads you got, which DJ had the most downloads. You kind of got points for that toward your account. That's one of my favorite websites. I also use a site called DJ City.com. This one is like a blogging website. We still share all different tracks but it is a website where guys will say, O.K., this is making its way down the East Coast. I've had stuff for four-six months before I started to hear it on the radio.

The site yourremix.com has closed down but DJcity is still a viable site to place mixes. DJ Merc likes both the ability to share his work and incorporate other DJ's mixes. Receiving statistics on his downloads and knowing who was most popular that month was important and it was one of his favorite sites. But what about sharing and incorporating digital music content?

> When I do a DJ gig, I get paid to go out there and do it, and I'm playing other peoples tracks with slightly different version of them with a different mix, but I didn't have the rights to that music. So it is kind of hard. It is really hard to say (DJ Merc).

The creation, the culture and the technology all combine to become complicit in the ethic questions concerning dubstep. The sampling of music a DJ has made is as easy as point and click. Adding beats, or as DJ Merc, explains, "a slightly different version with a different mix," allows the ownership water to become muddy. Sharing websites and the feedback and statistical "hit-making" increase the loop of sharing. The repeat ease of uploading and downloading by DJs and fans removes the "original" creator further and further from the mix (literally).

This becomes an ethical dilemma that the industry addresses in a variety of different ways. Most sites have copyright clauses and dubstep mixes are often flagged for copyrighted content and blocked from the site. The copyright infringement software for sites like SoundCloud and Youtube have become increasingly sophisticated and the algorithms can "catch" similar beats and times and identify hooks and mixes with copyrighted material in it. TIMarbury explains the impact collaboration has had for him:

> What was perfect about Newgrounds is anyone could post original music, so you could be really bad and post things there to get feedback on improving. When I first started, I was that kind of person. In 2008, I made about 0-5 tracks and I didn't really know if they were good enough to put online, so I released one track in 2009 and I got a positive response so I posted a few more and I started to get like 500- ,000 listens with 00+ downloads. It sounds silly but I was excited about this accomplishment.

Newgrounds has a page on their site called, "Collabinator, where a DJ can list his/her talent (an artist, a programmer, a musician, a voice actor, a writer, everyone) who is seeking (the same) as a place to connect similar to the 'wanted ads.'" Members also list competition information.

DJ Merc:

> It is hard to hear anything that is 100% original. In all honest, there is an origin in everything. So even if I go out and make a 100% original track, it is tough to say that it is mine because I listened to other people, and I put it together to make it my own. So I could be sampling from six different artists, and put it into one, and you could say that it is 100% yours because it was your idea, its your mix, but it was not 100% yours because you were clearly inspired by what other people are doing.

Sharing also leads to oversharing. Copyright issues have long been a concern in the digital music business.

DJ Merc:

> That is tricky. Some of the big players might [care]. That is kind of hard to say because it goes back again to the sampling. Whenever I think of copyrights I always think of the Vanilla Ice song that came out, the *Under Pressure* song just made that little sample. But with copyrights it is hard to say. With all the DJ sites out there, it is something that we go into, and we all know, well, if it is an original mix, I know that someone can take it. But that is a risk I take when I log onto that website. But then again it is not really a copyright loss for me because I did not get the rights to the track I am using. So I don't own it.

Most DJ sharing sites have a section that discusses copyright issues and several note how to get around copyright concerns. A subscription fee is part of most sharing sites, which adds a layer of protection on the idea of membership. SoundCloud's site give detailed information on the specifics of copyright and allows users to quickly be able to report any copyright infringement. The site provides a checklist and suggests DJ's ask themselves the following questions to specifically orient the site user to copyright concerns. The questions focus of whether or not the DJ composed and wrote the content. Further, the site specifically asks, "Does the track contain the entirety or any part of someone else's song(s). Is it based on someone else's song(s)?" Because, as the site says, "Copyright is complicated." DJcity takes a different route with their site. They explicitly note that the content providers retain the copyright to their content and DJcity is an independent subcontractor to promote the content.

THE MASHUP

Since my initial interest in 2011, Dubstep has ascended the ranks of popularity and strains of the dubstep nation can be heard, wub, wub, wubbing its way through popular culture in music, movies, commercials, and the "net." In the summer of 2012 *Spin* came out with the "Thirty Greatest Dubstep Hits of All time." Clearly, dubstep has arrived. TIMarbury's advice to others shares the foundation of the movement, have fun, don't be afraid, and be creative. And his # 1 tip?

> My number one tip is don't think you need to spend a lot of money or have the state of the art equipment to DJ or produce music, because what's going to make you successful is your talent, not the stuff you use.

Clearly, technology is central to the experience, though the human side of the experience, the talent, is still more important than "the stuff you use." This reflection has only scratched the surface of understanding the human-technology-world relation of the dubstep experience and the postphenomenological possibilities and variations have not been exhausted. There is still more to hear.

> Learning to hear the unsaid gains entry into this community and history to some degree. The learned is the initiate who has already heard and thus has entered into the community and the history. . .[however] the unsaid can be missing in unlearned listening. (Ihde 2007 62–63)

NOTE

1. DJ Merc is the professional name of Tim Mercandetti. He was interviewed for this chapter on March 20, 2012. Portions of his transcribed interview are shared here and he signed a permission for use in this publication.

REFERENCES

Cage, John. "The Future of Music." *Philosophy of Media Sounds*. Edited by Michael Schmidt. New York: Atropos Press.

Dekker, Annet, "Synaesthetic Performance In The Club Scene." *Cosign 2003: Computational Semiotics Proceedings*. August 19, 2011. *http://citeseerx.ist.psu.edu/viewdoc/download?doi=10.1.1.114.5725&rep=rep1&type=pdf.*

"Dodge and Fuski Interview." Dubstep 101.com note. August 19, 2011. http://dubstep101.com/photo/dodge-and-fuski-interview/.

"TIMarbury Interview." Dubstep 101.com note. August 19, 2011. http://dubstep 101.com/photo/timarbury-interview/.

Herrera, Monica. "Skrillex Isn't Surprised By Dubstep Takeoff." *Rolling Stone Magazine*. November 2, 2011.

Ihde, Don. *Experimental Phenomenology*, Albany: State University of New York Press. 1986.

Ihde, Don. Postphenomenology and Technoscience: The Peking University Lectures. 2009.

Ihde, Don. *Listening and Voice: Phenomenologies of Sound*. Albany: State University of New York Press. 2006.

Ihde, Don. *Postphenomenology: Essays in the Postmodern Context*. Evanston: Northwestern University Press. 1993.

Ihde, Don. *Experimental Phenomenology*. Albany: State University of New York Press. 1986.

Irwin, Stacey. Interview with Tim Mercandetti. Personal interview. Lancaster, March 6, 2012.

Pfadenhauer, Michaela. "The Lord of the Loops. Observations at the Club Culture DJ-Desk." *Qualitative Social Research*. Sep2009. Vol. Issue 3, pp. 11–17.

"The Thirty Greatest Dubstep Songs of All Time," *Spin*. July 24, 2012. http://www.spin.com/articles/making-brand-40-greatest-band-names-all-time/.

vanVeen, TC, Bernado A Attias. "Off the Record: Turntablism and Controllerism in the 21st Century (Part 1)." *Dancecult: Journal of Electronic Dance Music* (2011): Vol. 3. Number 1. DOI: 10.12801/1947–5403.2011.03.01.08.

Verbeek, Peter-Paul. *What Thing Do: Philosophical Reflections on Technology, Agency, and Design*. Translated by Bob Crease. University Park: The Pennsylvania State University Press. 2005.

Chapter 9

Postphenomenology

What's New? What's Next?

Robert Rosenberger

I would like to take the opportunity to offer encouragement to those working on postphenomenology to develop creative expansions of this theoretical perspective. Insofar as postphenomenology is already a blooming school of thought, much of this development has thus far come in terms of the application of postphenomenological ideas to concrete case studies of human-technology relations. I argue here that the time has come to take the next generation of postphenomenological insights, specifically those emerging from these cases, and use them in the analysis and expansion of postphenomenological theory.

By the school of thought of "postphenomenology" I refer to the interdisciplinary and international group of researchers studying human relations to technology through a shared general collection of philosophical insights, concepts, and commitments (see, e.g., Ihde, 2009; Verbeek, 2011; Rosenberger & Verbeek, 2015). One key commitment is to the integration of a conception of human experience developed in the philosophical tradition of phenomenology with the ontological and epistemological claims made in the tradition of American pragmatism. This perspective has been first and primarily developed in the career-long body of work of the philosopher Don Ihde. And typical work in postphenomenology proceeds through the application of Ihde's philosophical framework of concepts to concrete cases studies of actual user relations to specific technologies, from laboratory instrumentation, to everyday tech such as phones and computing, to design practices. (In addition, some of this development has occurred through the exploration of the places of overlap with other perspectives in the philosophy of technology and STS, including Heidegger studies, ANT, and the works of Albert Borgmann, Andrew Feenberg, Donna Haraway, and others.)

For example, there is a burgeoning body of postphenomenological work on laboratory and medical image interpretation (e.g., Ihde, 1998; Hasse, 2008; Carusi, 2009; Rosenberger, 2011a; Verbeek, 2011; Forss, 2012; Friis, 2012; Rosenberger, 2013a; Carusi and Hoel, 2014; Friis, 2015; Hasse, forthcoming). In these studies, images are explored in terms of the perceptual relationships users develop with them through their interpretation. As Ihde puts in *Expanding Hermeneutics*, his path-clearing work on this line of investigation, "The role of repeatable, Gestalt patterns . . . is the epistemological product of this part of the quest for knowledge" (1998, 171). Anthropologist Cathrine Hasse uses these ideas to explore training practices for image interpretation in physics (e.g., 2008; forthcoming). Anette Forss uses these ideas as part of ethnographies of cancer screenings (e.g., 2012). And I have analyzed ongoing debates over image interpretation in neurobiology and space science (e.g., Rosenberger, 2011a, 2013a). This imaging work—just one of postphenomenology's topic areas—shows the empirical and practical character of this school of thought, as well as its interdisciplinarity, stretching across not only philosophy and the target areas of science and medicine, but also anthropology, sociology, human-computer interaction, media studies, women's studies, and STS.

But as postphenomeonlogy's catalog of case studies grows richer, new tasks are emerging. It is time to begin capitalizing on these very cases as tools for interrogating the received postphenomenological framework of ideas. That is, it is time to identify places in these studies in which established postphenomenological ideas strain to give an adequate account. It is time to chart out the boundaries and limitations of these ideas. And it is time to expand and refine these ideas, addressing those limitations, and pressing those boundaries. This project will involve any number of tasks, from modifying existing ideas, to developing new concepts, to synthesizing the postphenomenological perspective with other theories.

This work has of course already begun. And here I would like to review what I take to be the cutting-edge efforts on two "pressure points" in postphenomenology: implanted technologies; and the implications of multistability for postphenomenological methodology.

The first issue regards technologies that a user has implanted into his or her own body, from pacemakers, to neuro-stimulators, to science fiction stories of bionic eyes or computer link-up ports. Such technologies have the potential to disrupt the typical postphenomenological conception of human-technology relations because this perspective so far has focused mainly on how users engage with devices through bodily interaction and perception. As we will see, it is not immediately clear how established postphenomenological concepts could even approach the issue of implantation. The path forward into this issue is already being tackled by thinkers such as Helena De Preester,

Peter-Paul Verbeek, and Kirk Besmer, with Besmer's analysis of cochlear implant technologies emerging as a guiding case study. In the first section of this chapter, I review and comment on this line of thought.

The second issue regards the key postphenomenological notion of multistability and its methodological implications for this perspective. Multistability refers to the capacity of any technology to support a variety of uses and meanings. This idea is often wielded by postphenomenologists against totalizing accounts of technology that fail to recognize the variability and context-dependency of human-technology relations. But an issue simmering just under the surface of much empirical work in this perspective is exactly how the notion of multistability should be used within concrete case studies. Methodological questions include: How, exactly, should the specific multiple stabilities of a technology be identified? Is there a difference between the use of the notion of multistability for critiquing others' claims, and its usage in the discovery of new information about our technological situation? What do our answers to these sorts of questions imply about the underlying epistemology of postphenomenology? Hasse, Kyle Powys Whyte, and I have been wrestling with these questions, and more and more it is appearing that their answers will involve the further development of postphenomenological epistemology and methodology. I review and expand on this issue in the second section of this chapter. These two budding discussions can serve as exemplars for the project of identifying and addressing the pressure points of the postphenomenological framework.

IMPLANTATION AND EMBODIMENT

One major pressure point emerging for the postphenomenological perspective regards the issue of implanted technologies, those devices which are integrated into the human body. The standard postphenomenological framework of concepts has not been designed with implants in mind; these concepts have been developed for describing the experience of bodily interaction with technologies, such as using a technology with our hands, wearing a technology, or perceiving and interpreting a technology's readout. But are these same concepts adequate for describing the experience of implanted devices? Or must fundamental changes be made to this framework to address the issue of implantation?

Embodying Implants

A central feature of the postphenomenological framework is Ihde's account of the different forms that human-technology relations can take. In order to help refine our ability to describe user experience, and also to make a point about the variability of these experiences, Ihde outlines four different ways that human bodies interact with technology: embodiment relations, hermeneutic relations, alterity relations, and background relations.

By the term "embodiment relations," Ihde refers to technologies that are taken into and extend our bodily awareness as they are used (1990, 73). Here Ihde has in mind classic examples from the history of phenomenology, such as Heidegger's hammer, and Merleau-Ponty's blind man's cane. In these examples, the device itself is not the focus of the user's bodily experience; instead the user experiences a transformed world *through* the device as it is used. A user experiences a transformed ability to act on the world through the hammer, and yet remains barely aware of his or her grip on the device as it is used, occupied instead by the work at hand. The blind man is not occupied by the feeling of the device in hand, but instead senses the world itself through the end of the cane. In Ihde's terminology, a user "embodies" the hammer or the cane, and does so with a high degree of "transparency." He describes the design and usage of technology in terms of a "double desire" with regard to the notion of transparency:

> It is the double desire that, on the one side, is a wish for total transparency, total embodiment, for the technology to truly 'become me' . . . The other side is the desire to have the power, the transformation that the technology makes available (1990, 75).

In this way, Ihde formalizes a foundational phenomenological insight—that is, that bodily awareness does not in any simple way coincide with the limits of our physical form—for the topic of human-technology relations.

But Ihde's account extends beyond these standard phenomenological insights and out into different forms of human-technology relationships. For example, in contrast to "embodiment relations," he uses the term "hermeneutic relations" to refer to those technologies that we perceive directly in their use (Ihde, 1990, 80). In cases of hermeneutic relations, if a user is schooled in how to "read" the display of such a device, then she or he acquires a technologically transformed accessed to the world. A wall clock is an example. If one already knows how to "tell time," then, as one looks at and interprets the hands of the clock, one receives information on the precise time of day. I've already mentioned above (and will review in more detail in part two of this chapter) a line of postphenomeonlogical case studies regarding hermeneutic relations, namely, those regarding scientific imaging technologies. Ihde calls a third form of human-technology relation "alterity relations" (1990, 97). In case of alterity relations, a user interacts with a device in a way that is similar to how she or he interacts with other people. An example is a dashboard-mounted GPS device that calls out driving directions in a pre-recorded human-like voice ("Turn left at the next intersection"). As computer interface schemes continue to advance, we may see an increase in the prevalence of alterity-style interactions, and a line of postphenomenological thought is developing to contemplate the issue (Irwin, 2006; Wellner, 2014b; Bottenberg, 2015). Lastly, Ihde uses the term "background relations" to refer to those technologies that make up our environment, but with which we are not at the moment interacting directly (1990, 108). For example, the running refrigerator or humming climate control system informs our surroundings and yet remains "off to the side" (Ihde, 1990, 109).

In all of these cases, the technology plays a "mediating" role in human experience, coming between the user and the world, and transforming them both in the process. That is, in some broad sense, a relationship is established between the device and the user's body. Ihde denotes technological mediation with the following simple equation: I – Technology – World (1990, 85). This enables him to note the different forms of human-technology relations with different tweaks to the equation, for example casting embodiment relations as: (I – Technology) – World. The parentheses are added to the equation to reflect the joined nature of the user and the technology together encountering the world through the embodiment relationship. Hermeneutic relations are instead denoted as: I – (Technology – World). In the case of all technological mediation, the technology's role is always "non-neutral," reshaping the world and the user as it is used. The usage of technology thus always involves trade-offs, and never simply grants the user some desired end without also introducing other changes to take into consideration.

Ihde explores the issue of implanted technologies in the chapter of his book *Ironic Technics* entitled "Aging—I Don't Want to be a Cyborg" (2008, ch. 3). He claims that implantations should be understood in terms of

> what I have called *embodiment relations*, which were experienced uses of technologies which remain detachable, but which in use are quasi-transparent, and not technologies taken literally into or inside my body. Yet, I will maintain that the desire remains applicable to *cyborg technologies* (2008, 32).

That is, in his view the existing postphenomenological framework should be able to capture much of our experience of these technologies. In particular, these devices should fall under the notions of embodiment, transparency, mediation, and trade-offs. Ihde substantiates his claims though a review of a number of examples, several with which he has personal experience, such as tooth crowns, a heart stent, and a detailed account of his triple bypass surgery. He is careful throughout to emphasize not only the experiential changes brought by these technologies (from the changes in energy levels, to the experience of a hospital stay), but also the side-effect trade-offs (including the potential for complications, and the increased risk of further ailments).

The Cyborg Turn

Contrary to Ihde's account, a line of thought is developing within postphenomenology which claims that addressing the issue of implantation requires a significant expansion of the postphenomenological framework (e.g., Welton, 2006; Verbeek, 2008, 2011; De Preester, 2011, 2012; Besmer, 2012; Dalibert, 2014). The trailblazing work on this topic is performed by Peter-Paul Verbeek. He contends that in order for postphenomenology to capture the experience and effects of implanted devices, we must expand on Ihde's quartet of forms of human-technology relations. We need a new understanding of "cyborg intentionality." He writes,

> While the four human-technology relations in Ihde's approach revolve around technologically *mediated* intentionality, in which both (mediated) human beings and (multistable) technological artifacts are constituted, the notion of cyborg intentionality articulates how human-technology relations can also take on a physical character, forming an actual amalgam of the human and the technological, as in the case when pieces of technology actually merge with the human body. (Verbeek, 2011, 145)

Thus, Verbeek adds "cyborg relations" to the list of the potential forms of human-technology relations to help distinguish those devices that are simply used from those which are instead integrated into the body to form a new hybrid entity. To distinguish this relation from others, he provides the formula: (Human/Technology)—World.

Kirk Besmer has developed a detailed postphenomenological case study of cochlear implants (2012). In the case of this example of implanted technology, a device is permanently attached to the cochlea, providing stimulation to the auditory nerve. This implant receives radio signals from an external microphone, first translated through a speech processor. Besmer claims that, "This is the 'merging' of human and machine that Peter-Paul Verbeek talks about," and that cochlear implants can best be understood in terms of the notion of cyborg relations (2012, 306). His study explores the ways that new users learn to hear through the device, how they become accustomed to its use, and how the device ultimately reshapes the world in which the user lives. Besmer even claims that the complexity of the device makes it not only an example of a cyborg relation, but one that challenges distinctions between embodiment and interpretation.

Helena De Preester is another central contributor to this discussion. But she remains critical of the move to place a key conceptual distinction between technologies that are internal to the human body and those that are external. In particular, De Preester argues that we cannot base a distinction on the relative transparency of implanted and non-implanted technologies, "since tool use shares this characteristic withdraw into the sensorimotor body" (2011, 124). De Preester claims that we should instead focus on the distinction between technologies that extend our bodily capacities and those that are incorporated into our own sense of body (whether those devices remain internal or external). And she argues that the basis for this distinction should be our sense of bodily ownership associated with technology usage. She writes, "Tool use induces changes in motor and sensory capacities, but not in bodily ownership. True incorporation, in contrast, involves changes in the feeling of bodily ownership itself" (De Preester, 2011, 123). With these distinctions drawn, she goes on to explore the differences in the bodily extension and incorporation associated with bodily prosthetics, devices that transform our perceptual capacities, and cognitive extension technology.

Against the Cyborg Turn

As a contribution to this ongoing discussion on the implications of implantation technology, I would like to offer some pushback against the cyborg turn. On the one hand, I certainly agree with the intuition that something crucially distinct is happening in the examples of implanted devices. There is an important difference between technologies that are used (be they embodied, or hermeneutically read, etc.), and technologies that are physically combined with our bodies. I agree that the conception of "human-technology relations" strains to account for the experience of technologies that somehow become a part of the human. But, on the other hand, when we begin our analysis from the lived experience of implanted technologies, I suggest that it is not clear how, and where, and in what way (and perhaps even whether at all) the line should be drawn between embodiment and implantation *phenomenologically speaking*. That is, from the perspective of a "god's eye view" down upon human-technology relations, it seems obvious that there are important differences between a technology used by a person through a bodily interface, and a technology that operates from inside the person's body. But from a phenomenological approach, those distinctions are neither always immediately present, nor are they always immediately relevant.

De Preester's point is crucial here and warrants repeating: both non-implanted embodiment relations and implanted "cyborg relations" can involve varying levels of transparency. We could contrast, for example, the experience of someone who is at the moment conspicuously aware of their cochlear implant, and in that same moment deeply unaware of the eyeglasses worn outside of the body on his or her face. So if the starting point is this everyday moment of the experience of these technologies, then it does not seem that the salient distinctive experiential features of these relations will be their statues inside and outside the body.

De Preester's response, as noted above, is not to draw a categorical distinction between inner and outer tech, but between those that are cognitively apprehended as "mine." This certainly seems to have purchase on these phenomena. But I am not yet certain what should be done with this suggestion from a phenomenological perspective. On the one hand, the experience of "ownership" will certainly sometimes explicitly accompany the experience of the relation with an implanted device. On the other, this shift to a distinction based on feelings of ownership may ultimately sidestep the issue of implantation altogether (her intention, of course), and perhaps in an unsatisfying manner. For example, even in cases in which a user remains explicitly aware of the device within his or her own body, there may be moments in which implanted devices are not met with conspicuous feelings of the device as part of one's self. In addition, I suspect that we could identify examples in which some degree of the experience of "ownership" accompanies the use of non-implanted technologies to which we share embodiment relations (Perhaps one example might be a feeling of being personally impacted when the car one is driving is bumped by something while in motion). So while I agree that De Preester has identified an important (and hitherto unformulated) feature of human-technology relations, I do not think that the notion of ownership ultimately enables us to completely articulate the experiences of implanted and/or "incorporated" devices.

One direction I would like to see this conversation take is toward a deeper phenomenological description of one's own body awareness. There would be of course many resources to draw on, from phenomenologies of the experience of internal organs (e.g., Leder, 1990), to accounts of body image and body schema (e.g., Gallagher and Cole, 1998; and especially Besmer, 2015), and feminist accounts of difference in bodily experience (e.g., Young, 1990; Grosz, 1995; Braidotti, 2011; see esp. Dalibert, 2014). This discussion could also benefit from deeper analysis of the qualities shared between the experience of implants and the experience of embodiment relations (and also, perhaps, background relations). Similarities and differences could be drawn between different cases across these forms, such as in terms of their potential for transparency, and for feelings of De Preester-style "ownership," but also learning and bodily acclimatization, and qualities such as "sedimentation" and "field composition" as I have developed elsewhere (e.g., Rosenberger, 2012, 2014a, 2014b).

Based only on informal and off-hand personal discussions, I have come to suspect that Ihde himself will eventually come down in agreement with the cyborg turn, adopting the idea that "cyborg relations" should be understood as a form of human-technology relations helpfully contrastable with embodiment, hermeneutic, and other relations. And I certainly want to argue here that the cyborg turn represents one of the most exciting pressure points in the postphenomenological literature. But, in part for the sake of keeping this conversation unsettled, I myself come down in defense of Ihde's position in *Ironic Technics* reviewed above. That is, I maintain that ultimately implantations should be understood as a subclass of embodiment relations, and not as a distinct and new relation. I offer this defense as a provocation to supporters of the cyborg turn, in attempt to further dial up the pressure on this key postphenomenological pressure point. We must remain open to the possibility, however counterintuitive, that it will be *less* useful, and *less* true to lived experience, to categorize together all those technologies that are fitted into our bodies. We must remain open, then, to the possibility that it may be more pragmatic, and more in tune with phenomenological inspection, that implanted devices should be understood in terms of already established notions, such as embodiment relations and background relations. Contrast, for example, Besmer's account of the cochlear implant that remains at times conspicuous and active, with Ihde's account of the heart stent that is so transparent that it is almost forgotten. It is possible that, at least in some relevant respects, it would be more revealing to understand the cochlear implant as embodied and the stent as backgrounded than it would be to group them both together based only on the fact that they each break the skin. . .

Put as a challenge, we need to answer the following: if we do not view the difference between implantations and embodiment relations from a god's eye view, then how can we define this distinction from within descriptions of experience? And if we start from accounts of relationality from a phenomenological stance, then must the line of distinction still fall between those technologies that break the skin and those that do not?

MULTISTABILITY AND METHOD

A central feature of the postphenomenological framework is the notion of "multistability." Put generally, this refers to the capacity of any technology to be used for multiple purposes, to find meaning in multiple contexts, and to support multiple relationships with users. This idea has proven useful within postphenomenological projects both for articulating cases in which technologies have come to be used in more than one way, and for critiquing accounts that fail to appreciate technology's variability. Case studies performed under the postphenomenological banner explore the multistability of every kind of technology, from emotionally assistive robots, to organ donation protocols, to subway benches (Hasse, 2013; Rosenfeld, 2015; Rosenberger, 2014b).

But questions are emerging over exactly how the notion of multistability should be used within concrete case studies, and what its epistemological implications may be. Thinkers such as Cathrine Hasse, Kyle Powys Whyte, and myself have been working to identify the methodological issues introduced by this idea, and to develop recommendations [see esp. Hasse (chapter 12), this volume; Whyte (chapter 5), this volume].

In what follows, I first briefly review the history of the notion of multistability. I next review my own distinction between the positive and negative uses of multistability, Whyte's notion of "pivots," and my notion of "relational strategies." These ideas are then clarified though a review of a series of studies in which the notion of multistability has proven useful: postphenomenological analyses of scientific and medical imaging practices. Finally, I discuss some of the methodological issues arising around this central pressure point of postphenomeonlogy.

The notion of multistability is first developed in Ihde's early work on human perceptual experience (1986). There he uses Husserlian variational theory to analyze the experience of a series of visual illusions, that is, drawings that can be perceived in more than one way. An example is the Necker cube, a drawing that can be perceived as a three-dimensional cube if interpreted one way, and as a different three-dimensional cube if interpreted another. Through Husserl-style variational analysis, Ihde brainstorms further coherent visual possibilities, such as a two-dimensional hexagon, a bug straddling a hexagonal hole, and a cut gem shape. Ihde's point is that each perceptual possibility becomes perceptually available as a viewer learns each associated interpretive story. But once the story is known and the viewer learns to see the new visually coherent interpretation, it becomes possible to apprehend the drawing in terms of that story all at once in a perceptual gestalt. In this way, the drawings can be understood as multistable, capable of supporting multiple perceptual possibilities.

It is only later in Ihde's career that the notion of multistability is applied to human-technology relations (e.g., 1990). The term comes to refer to the way a technology can always have more uses than only that usage for which it was designed, and can always find meaning within contexts other than that which its designer had in mind. In this terminology, a technology is multistable, and thus can always support multiple "stabilities" or "variations" (the two terms are used interchangeably in this literature). And although always multiple, the number of stabilities open to any technology also always remains finite, limited by the device's concrete materiality to only those relations that are experientially stable. A simple example is the multistability of an automobile. A car is designed and manufactured to be a mode of transportation, and this could be understood as the car's "dominant" stability. But other stabilities can be seen in practice: a car can find new meaning as a collector's item; a car becomes a sporting instrument when used in a drag race; a car can be used as a temporary home, as when someone is "living out of their car"; a car is classified by law as deadly weapon; and of course many other stabilities can be identified or imagined. Yet at the same time those stabilities are not unlimited. A car simply cannot be put to just any purpose, nor can it find meaning in just any context.

Within the postphenomenological framework, committed as this perspective is to pragmatic nonfoundationalism, the method of variational analysis takes on a distinct role. Where for Husserl, the analysis of an object from multiple brainstormed perspectives has the effect of revealing that object's "essence," for Ihde this method and its objectives are entirely recast. He writes that for Husserl,

> Variations could be used to determine what was variant and invariant. I have also found this technique invaluable in any phenomenological analysis—but as I used this technique, I discovered something other than Husserlian "essences" as results. What emerged or "showed itself" was the complicated structure of *multistability* (2009, 12).

For postphenomenology, the outcome of variational analysis upon a given technology is the demonstration of that technology's status *as* multistable.

I find it helpful to divide the various uses that the notion of multistability has found within postphenomenological projects into two categories: positive and negative. In its negative usage, the method of variational analysis is put to work in the critique of some other claim (or philosophical perspective) that is allegedly totalizing, or essentializing, or in some way over-generalizing. That is, if someone claims that "technology X must always be Y," then a way to criticize this claim is to show that the technology is in fact multistable. Just a few examples of the negative use of multistability include Ihde's critique of Heidegger's hammer analysis (e.g., 1993; 1999); Ihde's critique of Heidegger's account of the typewriter (2010); and Verbeek's critique of Borgmann (2005). Galit Wellner's critique of my own account of the driver distraction of cell phones can also be interpreted as a negative use of multistability (Wellner, 2014a). (It is a critique with which I, of course, take issue. See, Rosenberger, 2014a.)

In the positive use of multistability, this notion is used to shed new light on a concrete case of human-technology relations. This can be done in a number of ways, such as through the identification of multiple stabilities that all actually occur in practice, or thought the consideration of imagined possibilities.

I suggest that a cutting-edge topic for postphenomenologcial thought—and a pressure point for research going forward—is to better understand the role of multistability within "positive" case studies. How should variational analysis be utilized in actual concrete cases? That is, how should it be taken up in studies in which the end result sought is something more than simply the demonstration that a technology happens to be multistable? From what sort of vantage point should postphenomenology be understood to operate if it at once purports to identify previously unnoticed stabilities, and at the same time makes no anchoring claims on a technology's essence? This is all to ask: what are the methodological and epistemological implications of the notion of multistability?

One clear step in the right direction is Kyle Powys Whyte's work on these questions (this volume). After reviewing several postphenomenological projects, he notes that the notion of multistability is used differently across them. For example, he contrasts Ihde's analysis of the history of archery with my own descriptions of human-computer interaction. Ihde conducts variational analysis of bow and arrow usage, showing this practice to be multistable, situated differently within different historical and cultural contexts (2009). For example, Ihde identifies the English longbow and Mongolian horse-mounted bow as different variations of the same general bow-and-arrow practice, set within different cultural circumstances, and involving different advantage/disadvantage trade-offs. [Elsewhere Ihde goes further and claims archery to be a variation of a wider "bow under tension" trajectory of technological development, one which includes additional variants such as string instruments and the use of a bow string twisted around a stick to facilitate fire-starting friction (Ihde, 2012)]. A second employment of multistability is exemplified by my own studies of computer interface in which the same computer is approached in different ways in different circumstances, such as when the computer is functioning normally, and when an internet webpage loads more slowly than anticipated (e.g., Rosenberger, 2009, 2013b). (Whyte uses my example of the multiple uses of a simple glass bottle, which I developed in these essays on computing.) Whyte claims that these two projects rest on different "pivot points." That is, they each have a different kind of target of investigation analyzed *as* multistable. Where Ihde explores a broad set of archery-related practices in terms of its multistability, I instead analyze an individual device as the target technology capable of supporting multiple relations. Whyte recommends that postphenomenologists be clearer and more explicit regarding the kinds of pivots upon which their investigations rest.

Another new concept related to the notion of multistability is what I call "relational strategies" (e.g., Rosenberger, 2009, 2013b, 2014b). This term refers to the particular bodily comportments, perceptual anticipations, understandings, and habits that enable a user to relate to a technology in terms of one stability rather than others. This notion is useful for emphasizing the roles played by the human side of human-technology relations. That is, while it remains true that any technology is capable of supporting multiple stable human-technology relationships, it is also true that any of these stabilities is simultaneously also made possible by the particular manner in which the human user engages that technology. If we return again to the simple example of the multistable car, we can note that it is not only the case that the car is capable of affording multiple uses and meanings, it is also the case that each stability of the car is made possible by a human user that approaches the car in the terms of that stability. For example, the dominant stability is the use of the car for normal driving. This of course additionally calls for a user that "knows how to drive," and that she or he actively embodies the car in this manner. A user that takes a nap in a car is instead someone who approaches the car in terms of a different relational strategy—a relational strategy which involves both an understanding of this alternative usage, and also the comportment of the body into a sleeping position.

Multistable Imaging Technologies

The above notions of pivot points, relational strategies, and the positive use of multistability can be instantiated with examples from one of postphenomenology's most prolific areas of study: imaging technologies. This collection of studies builds on the "minimonograph" of insights into the phenomenology and hermeneutics of image use in the laboratory included in Ihde's 1998 book, *Expanding Hermeneutics* (see, e.g., Hasse, 2007; Verbeek, 2007; Carusi, 2009; Rosenberger, 2011a, 2011b, 2013a; Forss, 2012; Friis, 2012, 2015; Carusi and Hoel, 2014; Hasse, forthcoming). Ihde's basic argument is that image interpretation in science should be understood in terms of the perceptual relationships that scientists establish with laboratory image-making devices. Image use should thus be conceptualized in terms of hermeneutic relations to technology. That is, as a user looks at and interprets the readout of a laboratory imaging device, she or he receives a transformed relation to the world. And if the user is already an expert at reading a particular type of image, and as such brings expectations and a sort of pre-perceptive readiness to the experience, then much of the meaning of the image will be perceived all at once in a gestalt.

Many postphenomenological studies on image interpretation in science and medicine are paradigmatic examples of the positive use of multistability. Take, for instance, the work of Cathrine Hasse (e.g., Hasse, 2007; Hasse, forthcoming). She studies the ways that images in the physics laboratory are experienced differently by those working in different countries, and by those of different generations. In his own work on the topic, Ihde has considered the multistability of imaging technologies across two very different domains. In the first line of thought, he considers the "instrumental trajectories" of scientific practice, that is, the long lines of developing laboratory technologies that guide scientific work. (This view, in part inspired by the later work of Heidegger, can be set against traditional accounts of science as primarily theory-driven.) Ihde explores the development of contemporary imaging technologies as one variation among several emerging within a visualist trajectory in science with roots in early telescopes and camera obscura devices (e.g., 1998; 2009; 2012). In a second line of thought, Ihde reflects on the possibilities for investigating the same object of study via different imaging devices, and he considers the implications for the philosophical issue of scientific realism (1998). Anette Forss studies the workplace practices of microscopists screening for cancer (1998). Forss explores the multistability of these samples in terms of the different interpretive approaches that the screeners can adopt, including one that takes into consideration the lives of the patients behind the samples. In my own work on this topic, I explore scientific debates over images, and I suggest that such debates should be reconceptualized in a language of multistability (e.g., Rosenberger, 2011a, 2011b, 2013a). I argue that such a reconceptualization helpfully highlights the relevant instrumental practices. My studies include disagreements over the nature of neurotransmission made possible by complex freezing techniques, and debates over the climatological history of Mars made possible by a continuing precession of satellite-mounted cameras. My suggestion here is that all of these postphenomenological cases are examples of the positive use of multistability; in different ways each of these studies investigates actual stabilities of multistable technologies, and applies insights to the concrete (if however multiple) human-technology relationships possible with laboratory imaging devices.

It is helpful to consider these studies in terms of the Whytian pivot points that inform each approach. That is, we can ask: what kind of target does each postphenomenololgical study of scientific imaging reviewed above investigate *as* multistable? As in the case of the contrast between the archery and computing examples above, we can identify an analogous general distinction emerging within these studies. Again it is a distinction between studies which target a cluster of related imaging practices and those studies which instead target a specific image or set of images as multistable. For example, Ihde's investigation into a large-scale imaging trajectory in science does not have a specific technology as a central pivot point; instead he identifies a general trend in visualist imaging practices which develops over time and yields an assortment of variations through the history of science. This is true as well, though in a different way, for Ihde's use of the notion of multistability in his thoughts on scientific realism; although it is at least purportedly the same invisible object of study at issue in his argument, that object is investigated through multiple and very different imaging systems. In these cases, Ihde's imaging studies share a general resemblance to his archery example.

In contrast, the postphenomenological studies on imaging performed by Hasse, Forss, and myself are of a different sort. These cases pivot on specific images. That is, specific images are investigated as multistable, and the different ways these images are interpreted in practice (or could be interpreted) are taken as different stabilities. For example, in Hasse's studies a set of images is investigated as multistable, and the different stabilities are constituted by the different meanings that are found within these same images by physicists from different national and generational contexts. In Forss's studies, the same samples are investigated as multistable, and the different stabilities are constituted by: (1) the typical interpretive practices, and (2) those enlarged practices for which she advocates. In my own studies, again the same images are investigated as multistable, and the different stabilities are constituted by the interpretations of these images offered by the opposing sides of an ongoing debate. In the end, these studies (by Hasse, Forss, and myself) in which the pivot is centered upon a specific set of images more closely resemble Ihde's Necker cube example than they do either his archery example or even, in a way, his own imaging studies.

The notion of "relational strategies" can be applied to these imaging studies as well. When utilized in this context, we can instead use the term "hermeneutic strategies," thus specifying that we are referring to a context of hermeneutic relations to imaging instrumentation. Ihde has used the term "hermeneutic strategy" in his work on the multistability of visual perception in general. He writes, "In a hermeneutic strategy, stories and names are used to create an immediate noetic context . . . The story creates a condition that immediately sediments the perceptual possibility" (1986, 88). I have developed this notion into a technical term for the study of scientific imaging through a series of papers (e.g., Rosenberger, 2011a, 2011b, 2013a). In this context, the term refers to the understanding and trained perception that enables the hermeneutic relation to a particular stability of a multistable image. In my own line of research I have been interested in ongoing debates over images, and the notion of hermeneutic strategies has been useful for unpacking the aspects of practice that make it possible for images to be the centering points of long-enduring scientific controversies. My observation is that often such debates are conceived by working scientists as disputes over reality itself; one's own side is correct about the world, and the other side is mistaken. I have urged that it is at times useful to reconceptualize these same debates as disputes over multistable images—that is, concrete material technologies open to multiple coherent perceptual relations. If a debate is tactically reconceived in this way, we are prompted to spell out and critically contrast what goes into the hermeneutic strategies that enable each of the rival image readings. Relevant features of hermeneutic strategies could include anything from theoretical underpinnings, to areas of focus within image content, to technical understandings of the image-making process (e.g., calibration, digital processing, "artifact" generation), to the roles of relevant non-imaged data. All of this kind of information and more goes into the trained pre-perceptions that enable a scientist to apprehend the image in terms of a particular stability, all within the context of a visual gestalt.

Implications for Methodology

As we begin to expand and refine our conceptualization of multistability, it is becoming clear that postphenomenology will need more explicit and more detailed methodological directives. When a postphenomenological project employs the positive usage of multistability, and is thus interested in more than simply showing *that* a technology is multistable, and aims to outline actual multiple stabilities, it will be important to lay out the methodological steps necessary for identifying and analyzing such stabilities.

The need for a refined postphenomenological methodology is not only practical, it is philosophical. That is, the issue is not simply that with a more nuanced understanding of multistability comes the corresponding need for a more nuanced method for making use of this notion in practice. I suggest that the expansion of the notion of multistability through notions such as "pivot points" and "relational strategies" introduces fundamental changes to this idea, and ones especially relevant to practice. They reveal the inextricable role of the postphenomenological investigator her or himself, and the role of the actual users of each known usage, in the investigation of multistable technologies.

If we halt our conception of technological multistability at the simple fact *that* technologies are multistable, then we are not compelled to address further issues regarding the subject position of the postphenomenological investigations, or those of the particular humans actually involved in the different human-technology relations constituting a technology's various stabilities. Or again, if we only make a negative argument criticizing someone else's failure to appreciate a technology's multistability, then we are only required to brainstorm some otherwise unnoticed stabilities, any stabilities. It does not matter how we come up with those stabilities, or which particular stabilities we offer.

But when we do more, when we conduct an empirical investigation of actual and potential stabilities, when we do an anthropology of technological multistability, we are engaging in a project with a fundamentally different epistemological character. If we explore the actual multiple stabilities of a technology, or the implications of particular stabilities, then subject positions matter. In such "positive" uses of multistability, inextricable hermeneutic issues arise. That is, inextricable questions arise regarding the postphenomenological investigator's relationship to the technology, and his or her relationship to the different potential users of the different potential stabilities under consideration.

This is all to say that an issue going forward for postphenomenology is to spell out the methodology and epistemology involved in the identification and analysis of the multiple stabilities of a target of investigation. This may end up requiring more than Ihde's own account of the creative brainstorming of phenomenologically trained investigators. It may require a more developed account of the postphenomenological investigator her or himself as one conducting the exploration from a particular subject position, one with only limited access to knowledge of only particular stabilities of a multistable technology. It thus may also require the articulation of the epistemological character of the kinds of claims that postphenomenology makes regarding technological multistability when such limited subject positions are taken into account. This could also require the development of this perspective's connections to feminist standpoint epistemology and Harawayan situated knowledge accounts, connections which have lingered on the periphery of postphenomenological thinking for decades, but which may now call for more central treatment.

CONCLUDING REMARKS

My goal above has been to identify and articulate what I take to be two of postphenomenology's most interesting and important pressure points, that is, places in which the conceptual framework of this perspective is beginning to buckle under the weight of the work being performed though this very perspective. While I hope to have made minor contributions to these two lines of thought—the discussions on implantation, and on multistability and method—I also hope that these ongoing discussions can be seen as exemplars. That is, I hope they can serve as guiding examples of how the cases studies performed through postphenomenology can be drawn upon as a resource for the task of challenging and expanding the postphenomenological theoretical framework.

REFERENCES

Besmer, K. (2012). "Embodying a Translation Technology: The Cochlear Implant and Cyborg Intentionality." *Techné: Research in Philosophy and Technology*. 16(3): 296–316.

Besmer, K. (2015). "What Robotic Re-embodiment Reveals About Virtual re-embodiment: A Note on the Extension Thesis." In R. Rosenberger & P-P. Verbeek (eds.), *Postphenomenological Investigations: Essays on Human-Technology Relations*. Lexington Books/Rowman Littlefield Press.

Bottenberg, F. (2015). "Searching for Alterity: What Can We Learn From Interviewing Humanoid Robots?" In R. Rosenberger & P-P. Verbeek (eds.), *Postphenomenological Investigations: Essays on Human-Technology Relations*. Lexington Books/Rowman Littlefield Press.

Braidotti, R. (2011). *Nomadic Subjects: Embodiment and Sexual Difference in Feminist Theory*. New York: Columbia University Press.

Carusi, A. (2009). "Philosophy Engines: Technology and Reading/Writing/Thinking Philosophy." *Discourse*. 8(3). http://www.prs.heacademy.ac.uk/view.html/PrsDiscourseArticles/69

Carusi, A. & A. S. Hoel. (2014). "Toward A New Ontology of Scientific Vision." In Coopmans, Vertesi, Lynch & Woolgar (eds.), *Representation In Scientific Practice Revisited*. Cambridge, Mass. & London: MIT Press 201–221.

Dalibert, L. (2014). "Posthumanism and Somatechnologies: Exploring the Intimate Relations Between Humans and Technologies." Thesis, University of Twente.

De Preester, H. (2011). "Technology and the Body: The (Im)Possibilities of Re-embodiment." *Foundations of Science*. 16: 119–137.

De Preester, H. (2012). "Technology and the Myth of the Natural Man." *Foundations of Science*. 17: 385–390.

Forss, A. (2012). "Cells and the (Imaginary) Patient: The Multistable Practitioner–Technology–Cell Interface in the Cytology Laboratory." *Medicine, Health Care, & Philosophy*. 15: 295–308.

Friis, J. K. B. O. (2012). "Interpreting the Visual." *Philosophy & Technology*. 25: 249–253.

Friis, J. K. B. O. (2015). "Towards a Hermeneutics of Unveiling." In R. Rosenberger & P-P. Verbeek (eds.), *Postphenomenological Investigations: Essays on Human-Technology Relations*. Lexington Books/Rowman Littlefield Press.

Gallahger S., and J. Cole. (1998). "Body Image and Body Schema in a Deafferented Subject." In D. Welton (ed.), *Body and Flesh: A Philosophical Reader*. Oxford: Blackwell, 131–147.

Grosz, E. (1994). *Volatile Bodies: Towards a Corporeal Feminism*. Bloomington: Indiana University Press.

Hasse, C. (2008). "Postphenomenology: Learning Cultural Perception in Science." *Human Studies*. 31(1), 43–61.

Hasse, C. (2013). "Artifacts That Talk: Mediating Technologies As Multistable Signs and Tools." *Subjectivity*. 6: 79–100.

Hasse, C. (forthcoming). *The Anthropology of Learning and Organizational Culture*. Springer.

Hasse, C. (this volume). "Multistable Roboethics." In J. K. B. O. Friis, & R. P. Crease (eds.), *Technoscience & Postphenomenology: the Manhattan Papers*. Lexington Books.

Ihde, D. (1986). *Experimental Pheneomenology, second printing*. Albany: SUNY Press.

Ihde, D. (1990). *Technology and the Lifeworld*. Bloomington: Indiana University Press.

Ihde, D. (1993). *Postphenomenology: Essays in the postmodern context*. Evanston, Il: Northwestern University Press.

Ihde, D. (1999). "Technology and Prognostic Predicaments." *AI & Society*. 13: 44–51.

Ihde, D. (2008). *Ironic Technics*. Automatic Press/VIP.

Ihde, D. (2009). *Postphenomenology and Technoscience: The Peking University Lectures*. Albany: SUNY Press.

Ihde, D. (2010). *Heidegger's Technologies*. New York: Fordham.

Ihde, D. (2012). *Experimental Phenomeonlogy: Multistabilities, second edition*. Albany: SUNY Press.

Irwin, S. (2006). "Technological Other/Quasi Other: Reflection on Lived Experience." *Human Studies*. 28: 453–467.

Leder, D. (1990). *The Absent Body*. Chicago: University of Chicago Press.

Rosenberger, R. (2009). "The Sudden Experience of the Computer." *AI & Society*. 24: 173–180.

Rosenberger, R. (2011a). "A Case Study in the Applied Philosophy of Imaging: The Synaptic Vesicle Debate." *Science, Technology, & Human Values*. 36(6): 6–32.

Rosenberger, R. (2011b). "A Phenomenology of Image Use in Science: Multistability and the Debate Over Martian Gully Deposits." *Techné: Research in Philosophy and Technology*. 15(2): 156–169.

Rosenberger, R. (2012). "Embodied Technology and the Problem of Using the Phone While Driving." *Phenomenology and the Cognitive Sciences*. 11(1): 79–94.

Rosenberger, R. (2013a). "Mediating Mars: Perceptual Experience and Scientific Imaging Technologies." *Foundations of Science*. 18: 75–91.

Rosenberger, R. (2013b). "The Importance of Generalized Bodily Habits for a Future World of Ubiquitous Computing." *AI & Society*. 28: 289–296.

Rosenberger, R. (2014a). "Multistability and the Agency of Mundane Artifacts: From Speed Bumps to Subway Benches." *Human Studies*. 37: 369–392.

Rosenberger, R. (2014b). "The Phenomenological Case for Stricter Regulation of Cell Phones and Driving." *Techné: Research in Philosophy & Technology*. Forthcoming.

Rosenberger, R. & P.-P. Verbeek. (2015). "A Field Guide to Postphenomenology." In R. Rosenberger & P.-P. Verbeek (eds.), *Postphenomenological Investigations: Essays in Human-Technology Relations*. Lexington Books/Rowman Littlefield Press.

Rosenfeld, A. (2015). "Mediating Multiplicity: Brain-Dead Bodies and Organ Transplant Protocols." In R. Rosenberger & P.-P. Verbeek (eds.), *Postphenomenological Investigations: Essays in Human-Technology Relations*. Lexington Books/ Rowman Littlefield Press.

Verbeek, P.-P. (2005). *What Things Do: Philosophical Reflections on Technology, Agency, and Design*. University Park: Penn State University Press.

Verbeek, P.-P. (2007). "Beyond the Human Eye: Technological Mediation and Posthuman Visions." In P. Kockelkoren (Ed.), *Mediated Vision*. Rotterdam: Veenman Publishers, pp. 43–53.

Verbeek, P.-P. (2008). "Cyborg Intentionality: Rethinking the Phenomenology of Human-Technology Relations." *Phenomenology and the Cognitive Sciences*. 7(3): 387–395.

Verbeek, P.-P. (2011). *Moralizing Technology: Understanding and Designing the Morality of Things*. Chicago: Chicago University Press.

Wellner, G. (2014a). "Multi-Attention and the Horcrux Logic: Justifications for Talking on the Phone While Driving." *Techné: Research in Philosophy & Technology*. Forthcoming.

Wellner, G. (2014b). "The Quasi-Face of the Cell Phone: Rethinking Alterity and Screens." *Human Studies*. 37: 299–316.

Welton, D. (2006). "Body and Machines." In E. Seligner (ed.), *Postphenomenology: A Critical Companion to Ihde*. Albany: State University of New York Press, 197–209.

Whyte, K. P. (this volume). "What is Multistability? A Theory of the Keystone Concept of Postphenomenological Research." In J. K. B. O. Friis & R. P. Crease (eds.), *Technoscience & Postphenomenology: The Manhattan Papers*. Lexington Books/ Rowman Littlefield Press.

Young, I. M. (1990). *Throwing Like a Girl and Other Essays in Feminism and Social Theory*. Bloomington: Indiana University Press.

Chapter 10

An Introduction to Hyperology

The Age of the Chimera

Roisin Lally

Given the current drive toward biotechnology and genetic engineering, the question concerning technology remains a leading question for philosophers today. The complex relationship between humans and technology has been widely documented,[1] yet contention lies within the conflicting critiques of technology. On either end of the spectrum are substantivism and instrumentalism (namely, the substantivist principle of an autonomous force underlying technology and the instrumentalist principle of human control over technology). Taken in the broadest sense, substantivism remains within the modernist tradition, most notably related to Martin Heidegger and Jacques Ellul, while the latter is identified with the postmodern tradition, commonly associated with Don Ihde, and John Dewey. Ihde argues that Heidegger does not overcome the essentialist project that seeks a stable condition of possibility in which to ground truth. He is critical of the work of phenomenology on these grounds, in particular Heidegger who, he argues, falls back into the illusion of the old metaphysics of presence. For Ihde because modern technologies are radically different to traditional technologies, they require a different philosophical methodology to understand them. He argues that the modernist project adheres to the illusion of an ultimate truth and an absolute language, which ultimately leads to the illusion of an essential reality. Rather, for Ihde, the world of contingent presentations proves to be more authentic and primary than the so-called true reality.

To make sense of these technological transformations, Ihde argues that we must make a methodological shift from the modernist phenomenological tradition to what he coins "postphenomenology." Thus, by naming postphenomenology, he constitutes it as a new cultural paradigm by differentiating itself from phenomenology; it is the difference between stability and multistablity (Ihde, Heidegger's Technologies, 2010) a difference, he argues, that is overlooked by both Husserl and Heidegger. However, the notion of multi-stablity is already at play in the work of Heidegger. It is understood in terms of art. Art is a movement; more specifically it is a movement of truth. Truth is not identified with essence or stability or adequation; rather truth constitutes a genuine active participation in the making of and working out of a culture.

The move to postphenomenology in Ihde is philosophically substantive, however. He is interested in shifting the philosophy of technology out of the realm of metaphysics and into the realm of experiential scientific research by analyzing specific technologies. The tension between these two thinkers, therefore, is not the search for something beyond phenomenology but in the philosophical interests that drive their use of this method and their interpretation of its results. Ihde claims postphenomenology is purely empirical while Heidegger claims it to be ontological.

The task of this essay is to explore the interrelatedness of modernity and postmodernity as presented by Heidegger and Ihde, under a general concept hyperology. By clarifying the concept "hyper" as a condition of possibility for human-existence, I argue that hyper, hitherto thought of as a tension between the amplification and the negation of reality, functions as first-order distantiation of the human being; it is a pre-reflective, basic condition of human existence. In this way it is similar to Ihde's embodiment theory that he outlines in (Ihde, Bodies in Technology, 2002). However, I depart from Ihde on his theory of the "hyperworld" where he contends the hyperworld is a projection into the future: a projection into virtual reality (VR) (ibid, xiv). For Ihde this can only happen through instruments. I argue, on the contrary, *hyper* is the projection of the self onto the world; it serves as the bridge between humanity and culture. It is the-there or in Ihde's term, the "over-there" of the "real body," oscillating between the future time to come and the past that has been. Through imagination, *hyper* brings things into appearance that is invisible. This present essay will focus on works of art in particular works of bioart by two contemporary artists, Eduardo Kac and Patricia Piccinini.

Hyperreality is a postmodern term used to signify the Information Age. Paradoxically, hyperreality has been described as a world that ranges from excessive reality to a nonexistent reality.[2] The ambiguity lies in our understanding, or lack thereof, concerning the drive toward a *hyper-existence.* Hyperreality is not an uncommon concept. Etymologically *hyper* is taken from the Greek word *hupèr* meaning "over," "above," "above measure," or signifying a condition above or beyond. It's opposite, *hypokeimenon*, was used by Aristotle to describe the natural world of essences. For the most part, hyper was lost to philosophical reflection until the latter part of the twentieth century when it took on a different meaning with the advent of the Internet. People began to refer to the language of the Internet as hypertext and hyperspace, which was understood as a negation of both word and space.

The concept hyperreality has been advance by the French sociologist and philosopher, Jean Baudrillard in his book *Simulation and Subterfuge* (Baudrillard, 1994). He describes it as a conceptual point at which reality becomes indistinguishable from simulation, implying a presence that is nonexistent. It is the disappearance of reality brought about by the dominance of the mass media, a concept that is heavily influenced by Sausserian linguistics, in which signs are perceived to be an arbitrary psychological union of a signifier (sound image) and the signified (concept) (William, 1996) and signs only convey meaning through their relative position to other signs (Saussure, 1959). In hyperreality the experiential aspect of the subject that exists as interplay between temporal reality and the internal world of myth/ideology is distorted by simulations, thus threatening to destabilize the border between the real and the imaginary. The simulation is the implied presence of something that is nonexistent, producing a hyperreal: "the product of an irradiating synthesis of combinatory models in a hyperspace without atmosphere."[3] For Baudrillard hyperreality, through simulation, is no longer that of a territory, a referential being or a substance, it is the generation, by models, of a real without origin or reality: a hyperreal. Baudrillard uses the example of Disney's Main Street as a kaleidoscope of hyperreal representations. It is, for Baudrillard, an orbital recurrence of models without reference to anything that is real, yet it emerges as a more authentic, exact, "real" than the reality that surrounds us.

Illustrative of this is hyperrealism, an art movement of the 1970s and early 1980s that includes the works of Andy Warhol. Here the real and the imaginary, production and art, are "confounded in the same operational totality."[4] The world of art no longer transforms everyday life. Instead it absorbs the most ordinary experiences, producing the image over and over. Warhol's art, iconic in its ability to reproduce without ceasing to be art, is so successful in combining the machine and the metaphor, that "Unreality no longer resides in the dream or fantasy, or in the beyond, but in the real's hallucinatory resemblance to itself" (Baudrillard, 1994). Like hyperreal art, architecture spirals upward simulating a representation of reality immanent in its repetition. Baudrillard calls hyperrealism an "allegory of death": it is the meticulous technical reproduction of the real where death itself is absorbed into the simulation. Virtual reality is the home of postmoderns, a world of eternal recurrence and immanent repetition. Ihde's contribution is to think through the way hyperreality is embodied in technology. As embodied beings, or beings with technology, these projected fantasies have become constitutive of consciousness; they have become the new metaphysical way of understanding the world.

Ihde gives a phenomenological account of the human body as embodied. Embodiment is, for Ihde, an unmediated perceptual-bodily experience. He conducts a hermeneutical phenomenology of embodied beings, and concludes that phenomenology is only appropriate when talking about an unreflective life *without* technology. As bodies in technology we no longer live in this purely phenomenological world. Particularly as we communicate or play in virtual reality, we are absolved from the limits of normal embodiment as subjects confronted by objects. In these experiences, things that would have once stood out before us as objects become part of our lived experience. In his examination of body identification he breaks down the body into two components: an active "here-body" [sight 1] and imaginative "over-there" body [sight 2]. In his analogy the over-there body constitutes the world of fantasy, or as he suggests, a hyperworld. Knowledge of the world begins with body one, and projects outward into body two, through instruments and imagination. In contrast to embodiment we experience disembodiment when we communicate or play in virtual reality. However, hyper-reality cannot stand alone; it can never take the place of "real life."

In the example of the telescope, the reflexive seen-seeing of the image body [sight 2] and the here-body [sight 1] become an embodied observation even though its observation is not direct but mediated instrumentally (Ihde, 2002, p. 48). Here we have both the observed [sight two] and the observer [sight one]. The object to behold, thus far, remains invisible; it lacks visibility until the perceiver embodies the telescope. The observer and the instrument engage in a hermeneutical style of envisioning phenomena.

Prosthesis offers another example of embodied technology. Such devices are experienced through the body; however, this is never a simple replacement for the mode of embodiment that existed prior to the technological adaptation. The equivocal character of the prosthesis works in two ways: one in partial concealment, the other in partial withdrawal. For example, in hot weather prosthesis might conceal the heat of the day, or the hot earth, where the wearer cannot *feel* the earth beneath her feet. On the other hand, she might be more sensitive to the conditions of icy weather, where the prosthesis is more likely to slip. Thus, Ihde argues, these technologies can never become a full simulacrum of bodily sensory experiences, even though they are technologically embodied. Prosthesis is an extension of the here-body; the here-body and over-there body are interdependent. This is correct, and the high visibility of prostheses in our culture is changing the way we understand our body and the relation to the world. But can it be said that these technologies are a radical break with the past? Prosthesis, according to Umberto Eco, is

> any artificial construction, which prolongs and amplifies the possibilities of our body, from the first sharpened flints through to the lever, the walking stick, the hammer . . . In this sense the term prosthesis also cover chairs, or beds or clothes . . . they are all but natural extensions, and like our body, we take care of them and decorate them. (Eco, 2004, pp. 382–383)

We are embodied beings prior to our postmodern condition, prior to the simulation of reality through repetition, and prior to any scientific orientation to the world. As such, we have not moved radically away from the modernist project. We take care of our extensions because they matter to us, and we decorate them because meaning and truth are found in beauty. Unlike Baudrillard's dystopian view of art as an "allegory for death," Eco, like Heidegger, experiences beauty in all things and so it is an allegory of truth. In the "Origin of the Work of Art" Heidegger writes: art "makes public something other than itself; it manifests something other; it is an allegory" (Heidegger, OWA, 2009, p. 145). What matters is not what shows itself but only the possibility of *something* showing itself. Interpreting Heidegger, Günter Figal states that art is not something new, but lets things as a whole be seen anew (Heidegger, 2009, p. 13). Art, or truth or beauty, is at the core of all extensions or technologies; this is just as true of things today as it was with, for example, the Greeks.

This play on the emergence of things can be understood by way of Heidegger's example of the Parthenon. The Parthenon, as an objectively present object, no longer signifies the truth of a culture, but for the Athenians it was symbolic of the preservation of truth for society. If we think of it in terms of the image body, the project of the fantasy or hyperworld is manifest in the Goddess Athena. The technofantasy, Athena, as the "embodiment" of truth lasted as long as the world projected itself in that particular way. The fact that the Parthenon no longer functions in this way does not eliminate the essence of truth; it merely means we shift our meanings from one technofantasy to another. Every projection is contingent and arises out of multiple possibilities. According to our analysis this is not essentialism. Indeed it seems as though the essence of hyper-existence functions as technofantasy; it comes to presence through the embodiment of man, machine, imagination, and life world, but is not reducible to these.

In the latter example, the form, Athena, follows from her function, a symbol of truth. For Ihde, the form, bionic man, shows up as a symbol of strength and speed. In both cases, hyper-existence functions as a way to project or map fantasies onto the world. This means that hyper-existence and its essence, technofantasy, is prior to technology. For Heidegger what is prior to technology is the movement at the core of art. He calls it aletheia.

In other words, hyper-existence is made possible by a first-order distantiation from the world. Creating a distance from nature, human beings make space for invention. Invention is a uniquely human characteristic which begins with the unconscious tinkering of available material. Etymologically, tinkering comes from the word "itinerant" (Lt: *mortālis* to have human origin and transient; subject to death, destined to die, temporary). In German, to tinker is *basteln* or to do handcrafts, to make, to build. In Irish, *tincéir* stems from the verb *ar tinneal*—"at-the-ready." Thus, to tinker constitutes a transient, knowing and know-how: to tinker means to craft. The tinker fixes in-place [*steln*] or dwells with his or her technical skills. The Greek word is *technē,* and is tied to the idea of mastery and apprenticeship. It is the repetition and practice of skill. So what is tinkering that makes it stand out in its own nature?

Martin Heidegger describes craft as the "strength and skill of the hands." The essence of hand is not reducible to brute fact, as a grasping organ. In this case apes, too, would be classed as humans because they can grasp, but apes do not have hands. On the contrary "Only a being who can speak, that is, think, can have hands and can be handy in achieving works of handicraft" (*ibid.*) The craft of the hand is not merely a grasping, catching, pushing, and pulling. As Heidegger writes:

> The hand reaches and extends, receives and welcomes - and not just things: the hand extends itself, and receives its own welcome in the hands of others. The hand holds. The hand carries. The hand designs and signs, presumably because man is assigned. Two hands fold into one, a gesture meant to carry man into the great oneness. The hand is all this and this is the true handicraft. (Heidegger, 1976, p. 16)

Seen in this light, crafting includes the crafting of ideas, rhetoric, building, and art, and is the fundamental nature of human beings.

In *Building Dwelling Thinking* Heidegger pursues this idea of building as integral to humanity. He traces the verb *bauen* back to *baun* which signifies dwelling. However, buan also means to build, which, he concludes, *is* to dwell (p.147). Furthermore, *bauen* in its original sense says *how far* the nature of dwelling reaches, to dwell nearby or a "near-dweller" (ibid), that is to dwell with our neighbors and also ourselves, because *bauen* is also the German word for *bin* as in *ich bin*, "I am." In all this we are embodied beings, not unlike Ihde's embodiment theory. For Heidegger, humans dwell not just as tinkers, but as preservers of both time and distance. Here building as preserving and nurturing is not necessarily making anything, in contrast to the simple structures of temple-building. The twofold nature of building is cultivation and construction. Similarly, Ihde understands technology as both construction and as an activity, both instrumentally and as meaningful. But whereas for Ihde, human beings construct their reality technologically, and then embody those technologies socially and culturally, wherein humans transform nature; for Heidegger, technology is not passive, it arises within a horizon of meaning. We could say technology *shows up* as a telescope, or prosthesis, or virtual reality. The difference is that for Ihde technology is apart from humanity until humanity embodies technologies, while for Heidegger, technology is coextensive with humanity.

Revisiting the example of Heidegger and Ihde, we could consider the Parthenon a mere thing, but that may infer just an aggregate of properties, which clearly it is not (Ihde, 2010, pp. 74–77). The essence of the temple, its *huper-keimenon* comes from the idea of the craftsman and shows up as a symbol of spirituality: Athena, for example, is symbolic of courage, strength, arts, crafts, and truth. What emerges is the essence of what already exists. In other words, the core of the temple is not merely the materials from which it was made or of Athena, but the truth that grounds and directs the materials (Heidegger, OWA, 2009, p. 149). The essence or core of the artifact comes to light through imagination or in Ihde's way the "image body" and not through the technology itself. From the material, using both imagination and tools, truth emerges in the form of beauty.

On the other spectrum we have transgenic art or bioart. In terms of the language of embodiment, bioart is more a revival than an innovation. Ihde writes: "it is a return to pre-modernity in the sets of cultural beliefs that things could actually transmute or metamorphose. Devils inhabiting human bodies, human witches taking on animal shapes, the possibilities of monsters, prodigies, and freaks—were pre-modern morphs" (Ihde, 2002, p. 5). This, Ihde reminds us, is evidenced in Plato's pre-cinema cave wherein images of images were the only "realities" for the dwellers prior to Platonic liberation of the cave into sunlight. For Ihde, these illusions are harmless so long as there is a distinction between theatre and daily life. But phenomenologically, Eduardo Kac has challenged this notion as genetic technology has made the fantasy of ancient mythology a multistable possibility (Kac, 2005).

Transgenic or bioart is a new art form based on the use of genetic engineering techniques to transfer synthetic genes to an organism or to transfer natural genetic material from one species into another, creating unique living beings. Molecular genetics allows the artist to engineer the plant and animal genome and create new life forms. *Alba* the "GFP Bunny" (Green Florescent Protein) created by Kac in 2000[5] was originally albino, now, when shined with blue light, Alba glows a bright green. Alba is specifically a transgenic art not created for the purpose of breeding. She is to be viewed as a piece of artwork, according to her creator. This seems to transcend the very notion of reality where distinctions between real and virtual, subject and object, art and science are broken down. They are in Ihde's words extended technologies, where beings are embodied with science and technology. On reflection, Kac's work is an art work only insofar as the telescope can be considered an art work. GFP functions to reveal proteins that are invisible to the naked eye, in the same way as the telescope also reveals what is always already there. Both technologies make the invisible visible. But as a complex social event, neither the telescope nor the rabbit reveal the mystery of life. If we fail to recognize the rabbit as existing for its own sake, by treating it as an object, we strip it of its very identity; it can neither exist in nature nor in the hyperworld of man.

Although an interesting scientific experiment in itself, it does not enter into the fray of beautiful chimeric artistry, envisioned by Kac, unless we are to objectify each living being as a work of art. Once it is exposed as bioart, the natural beauty of the rabbit disappears, and is replaced by a gimmick, that is neither beautiful nor useful. It is neither art nor technology. However, Kac's work anticipated the later works of bioart. It reveals the contemporary age of genetic engineering. Patricia Piccinini's work frames the bioscientific practices of manipulation and alteration of living beings. In her work she urges us to think of our new creations as blurring the distinction between human and nonhuman, between the organic and technological. Creating "new worlds" through stem cell research, genetic engineering, bioelectronics is already part of our world. She writes:

> The possibilities for my creations are already amongst us, and before too long the things themselves could turn up unannounced, without our ever having had the opportunity to wonder how much we want them. (Piccinini & Orgaz, 2007)

"Big Mother" from *Evolution* (Piccinini, 2005) displays a cultural world for understanding bioengineering and the possibilities for what is come. It highlights the similarities between humans and animals. She writes "One of my interests in acknowledging our animalness is also about trying to understand our humanity. There has to be more to life than genetics and biology." What does come to light is the crossing or transcending from human to nonhuman. Her work is a porthole into the past and future, oscillating between a chimeric ideal and a living reality. The distinction is so blurred that the boundaries no longer exist. The world of the chimera is a manifestation of the ambiguous relationship between the world of nature and the hypernatural world of man. This ambiguity is covered over when we think of ourselves as simply natural, or simply instrumental. Kac, Piccinini, and Ihde blur the distinction between science and technology, human and nonhuman. This blurring of boundaries can be interpreted reductively, but hyperology allows us to understand these blurrings in ways that preserve the ambiguous tension between science and technology, human and nonhuman. Bioart forces us to recognize the degree to which we are incorporating nonhuman beings into culture, and to take responsibility for the consequences.

Donna Haraway summarizes Piccinini's work as revealing that we must "learn how to confront the complexities of the new world in order to be able to move toward multi-species reconciliation." As Haraway says: "Apocalypse looms; in that story the past-nature is the time outside time and must be restored in all its innocence. That kind of time is utterly wild, i.e., outside the care of responsible generations." Piccinini's art predicts a future of transspecies, hyperbeings; beings that were once merely virtual reality. These beings are chimeras. As we enter into a face-to-face relation with these new beings, Piccinini urges us to cultivate practices of care. What this means for philosophical thinking is that the hyperworld is not just the "image body" or VR as Ihde would have it; it comes to presence even when we think it is illusionary, or nonexistent or indeed invisible: in other words, even when we are denying its very presence. Piccinin's work is a study of human beings as embodied in science and technology; a constructed world that hovers above the world of essences, a place where humans dwell to escape the wild-country.

For Ihde the human body has become the site of continuous transformation through our scientific orientation toward the world. While Ihde sees this as a radical break with traditional phenomenology I have argued that this is also seen in Heidegger's analysis of art as a movement. Art serves as the truth of a culture but is never a stable quality. If this is the case, postphenomenology still acts within the phenomenological tradition. What the proliferation of contemporary technology reveals is not the reduction of meaning to the contingent materiality of multiple embodiments, but the timeless truth that meaning has always emerged in the space that humans create in distancing themselves from nature, that is, in the hyperworld. Hyperreality is already part of structures of the mind; from where the event begins its projection, allowing things to come into view. Projection is a thrownness into the world, not in the postmodern understanding as an "overthrowing" or negation of the world. To be thrown means existence is always "out there" prior to its arrival in the world.

So we ought to understand Ihde's work on embodiment in light of hyperexistence as pre-reflective, first-order distantiations. However, unlike Ihde I would argue, with Heidegger, that as we dwell in the world as mortals, we cultivate and construct the world around us, a world that begins with the "image-body" or the hyperworld of imagination. Technology as art—as the very ground of possibility for humanity—brings humans from mere existence into a space for dwelling. In dwelling humans necessarily engage in a hyperexistence. If this is true, the only way to understand contemporary technology and indeed the very relation between technology, art, and philosophy is to take seriously the drive toward hyperreality, a study I call hyperology, a study toward which all the present reflections point us. In an age when we reach into the mystery of life, into the secrets of our genes to create "beautiful Chimeras" the danger is that we become a species in wild country, where we will once again have to survive, instead of freeing a space for living as enlightened beings.

NOTES

1. See Andrew Feenberg, *Questioning Technology*, for a detailed description of his four theories of technology including critical theory and instrumentalization.

2. See Borgmann, 1992, Epstein, 1999, Baudrillard, 1985, Dreyfus, 1995.

3. See also the hyperrealism of simulation is expressed everywhere by the real's striking resemblance to itself in Albert Borgmann's *Across the Postmodern Divide* (1994) which claims that new technologies are taking us into the sphere of hyperreality, a term he borrows from Baudrillard. He argues we are losing touch with our bodies, with nature, with other people and with focal things and practices.

4. Baudrillard, Jean, "The Critique of Originality," in *Jean Baudrillard: Selected Writings*, (ed.) Mark Poster. (Stanford: 1988), pp. 143–7.

5. Alba was created with GFP, an enhanced version of the gene found in Aequorea Victoria jellyfish first discovered by Osama Shimomura. This particular protein has the unique property of biofluorescence. In biofluorescence, available light is absorbed and converted into light of another wavelength, producing a different colored light. In the case of GFP, photons of light in the blue wavelength are converted into photons in the green wavelength. When introduced to mammalian cells, GFP produces fluorescence two times brighter than the original jellyfish gene.

REFERENCES

Baudrillard, J. (1994). *Simulation and Subterfuge.* (S. Glaser, Trans.) University of Michigan Press.

Eco, U. (2004). *History of Beauty.* (McEwen, Trans.). New York: Rizzoli.

Ellul, J. (1964). *The Technological Society.* New York: Vintage Books.

Feenberg, A. (1999). *Questioning Technology.* London, New York: Routledge.

Heidegger, M. (1976). *What is Called Thinking?* (Vol. XXI). (J. Glenn Grey, Trans.) New York: Perennial. Originally published by Max Niemeyer Verlag, Tuebingen under the title Was Heisst Denken? English translation by J. Glenn Gray 1954.

Heidegger, M. (1977 a). *The Question Concerning Technology.*

Heidegger, M. (1977). The Origin of the Work of Art. In M. Heidegger, & D. F. Krell (Ed.), *Basic Writings* (2nd ed., pp. 139–212). London, New York: Routledge Taylor & Francis Group.

Heidegger, M. (1993 a). *Basic Writings.* San Francisco: Harper.

Heidegger, M. (1993). The Question Concerning Technology. In M. Heidegger, & D. F. Krell (Ed.), *Basic Writings* (2nd ed., pp. 307–342). London, New York: Routledge.

Heidegger, M. (2009). On the Origin of the Work of Art: First Version. In M. Heidegger, *The Heidegger Reader* (pp. 130–150). Bloomington, Indianapolis: Indiana University Press.

Heidegger, M. (2009). *The Heidegger Reader.* (G. Figal, Ed., & J. Veith, Trans.) Bloomington, Indianapolis: Indiana University Press.

Ihde, D. (2002). *Bodies in Technology.*

Ihde, D. (2010). *Heidegger's Technologies Postphenomenological Perspectives.* New York: Fordham University Press.

Kac, E. (2005). *Telepresence & Bio Art Networking Humans, Rabbits, & Robots.* The University of Michigan Press.

Piccinini, P. (2005). Elizabeth A. Sackler Center for Feminist Art. *Feminist Art Base.* Melbourne, Victoria. Retrieved from http://www.brooklynmuseum.org/eascfa/feminist_art_base/gallery/piccinini.php?i=1002.

Piccinini, P., & Orgaz, L. F. (2007). The Naturally Artificial World. *Tender Creatures*

Saussure, F. (1959). *A Course in general Linguistics.* (W. Baskins, Trans.). New York: The Philosophical Library.

Schmidt, D. J. (2013). *Between Word and Image.* Bloomington, Indiana: Indiana University Press.

Selinger, E. (Ed.). (2006). *Postphenomenology A Critical Companion to Ihde.* New york: SUNY.

William, B. (1996). *The Simulation of Surveillance: Hypercontrol in Telematic Societies.* New York: Cambridge University Press.

Chapter 11

Nursing's Nightingale Needed a Lamp!

Anette Forss

The title of my essay is an intended echo of Don Ihde's paper entitled "Husserl's Galileo needed a telescope!" (Ihde 2011). The graphic image of Galileo needing a telescope encapsulates an irony regarding material matters in nursing. I want to pursue this irony by engaging with the iconic image, and trope, of Florence Nightingale and the technology of the lamp. Most people are perhaps primarily familiar with nursing as a regulated profession and practice, the modern origins of which are often symbolized by Florence Nightingale holding an oil lamp in her hand, leaning over a patient in a long dark corridor of tents of wounded soldiers in the Crimea, which is why she was referred to in the Victorian press as "the Lady with the Lamp."

Nursing as it exists today is a discipline that emerged during the second half of the twentieth century, during which time the lamp in its original form has been left long behind. But the changing lamp is always present; it is omnipresent in its shift through various technological forms. It is in this sense that, through the interchangeable concepts of materiality and technology, I offer a different interpretation of the lamp, informed by Ihde's work.

At first blush, the title of Ihde's paper "Husserl's Galileo needed a telescope!" seems to "fit" a common view in nursing; that is, that humans (nurses) use (neutral) technology for specific purposes—technology is comprised of "mute objects" in the hands of the nurse's intentional use. This view of human technology relations is, on the contrary, confronted throughout Ihde's work, in which he argues that Husserl "got it wrong" (Ihde 2011, p. 75) when using Galileo as an example for a critique of early modern science, because "Husserl's Galileo lacks the very mediating technologies which made his [Galileo] new world possible" (a.a. p. 78).

> If science is that human practice which yields new knowledge and knowledge which exceeds the bare or unaided senses, then only the science which is embodied in instruments which amplify or magnify ordinary capacities will qualify and that is the science produced by Galileo-with-his-telescope. (Ihde 2011, p. 77)

Similarly, the lamp in its various technological forms amplifies the seeing of depths and surfaces impossible to see by the naked eye in the inextricably related disciplines of medicine and nursing. Following Ihde, the lamp would be an example of technologically mediated perception. It is important to note here that "the lamp" signifies many technological attributes in which the key interaction between the nurse and the machine is that of illumination. In nursing and medicine these days, there is a symbolic connection between Galileo's telescope and Nightingale's lamp. We see this in the microscope, the endoscope, the scan, the X-Ray. All of these connect with the lamp—the light, the camera. The nurse and the patient are wrapped in technology, but not as capture, rather in an ontological moment that changes everything.

To continue on this theme, and importantly, according to Ihde, when the I-world relation is changed to an I-technology-world relation "this is more than a formal change; the way the world is experienced is changed ontologically" (Ihde 1990, p. 47). Putting this historically, the difference between "Nightingale used a lamp to see her patients in the dark," on the one hand, and, on the other, "Nightingale-with-her-lamp" is a difference between, in the former case, conceiving technology as neutral as opposed to the latter case as having and ontological "position." This is why seeing technology as neutral, or worse, forgetting technology, may have important consequences, not only in philosophical analyses of science but also in nursing philosophy itself. This is because philosophers of technology put technology and materiality at the center of philosophical inquiry, problematizing philosophy for privileging "ideas," while nursing scholars abandon technology and primarily cherish philosophy for its ideational qualities. This is the irony of which I spoke above. Let me unfold it further.

I would like to ask the following questions on technology in relation to nursing as a discipline. First, does forgetting technology and materiality as worthy of philosophical exploration mean unwittingly reinforcing an ontological division between subject and object, and between theory and philosophy, a division we seek to overcome, albeit by an "ideational detour" (arguing that nursing practice is/should be philosophically, theoretically, ethically and evidence based as in scientifically *informed*)? The second is rather a statement, the "lamp" reveals to us that, in practice, technology and nursing have always been intermingled. Third, is there a "double standard" for technology in nursing as a discipline, and does this mean that nursing is a modern or a non-modern project in a Latourean sense? In an oft-cited statement Latour

(1993, p. 138) claims that "modern humanists are reductionists because they seek to attribute action to a small number of powers, leaving the rest of the world with nothing but simple mute forces." According to Latour (1993), modernity involves a separation between human and things and between "two sets of entirely different practices" (a.a. p. 10) and as long as these are kept separate "we are truly modern" (a.a. p. 11). Here is Latour on the matter.

> The first set of practices, by "translation," creates mixture between entirely new types of beings, hybrids of nature and culture. The second, by "purification," creates two entirely distinct ontological zones: that of human beings on the one hand, that of nonhumans of the other. (Latour 1993, p. 19)

So, both practices are at work, and "as soon as we direct our attention simultaneously to the work of purification and the work of hybridization, we immediately stop being modern" (Latour, p. 11). To me, Latour is relevant for nursing in the sense that, as a discipline, nursing primarily portrays human and things as "two entirely distinct ontological zones" (Latour 1993, p. 19), while in practice the telescope and the lamp show us that such a radical separation is impossible. Although for example Birkler (2007, p. 125, my translation) acknowledges a certain degree of intermingling between "naturalistic and humanistic elements" technology is typically located to the former, as a thing.

So, it seems to me that nursing needs to address technology and materiality in terms of two sets of practices at work simultaneously. The first is conceiving humans and technology as entangled (in an Ihdean sense) and the creation hybrids and networks (in an Latourean sense). The second is conceiving humans and things as two different zones. According to Latour (1993, p. 140) the latter processes of "purification," is what makes large-scale experiments possible.

I will use my PhD, to give an example of how easy it is to relegate technology to the background although it is at the forefront of the inquiry. As PhD student I was part of a large-scale research project aimed at investigating a population based screening program (cervical cancer screening) in urban Sweden (Tishelman et al. 1999). The ambitious research design crafted by my supervisor(s) meant, in short, to move beyond traditional evaluation research on the basis that such research need to include different stakeholders' perspectives and a variety of methodological approaches (e.g. both qualitative and quantitative methods). Consequently this large-scale study was comprised of a series of sub-studies on different groups of people involved in the screening program, that is, gynecologists, midwifes, healthy women, women at risk, and immigrant women, to explore their experiences on cancer prevention and the program. Forgotten was the screening test itself, around which an entire screening program was built after it was brought to Sweden in the

1950s from the US by a young Swedish "preparatris" (laboratory personnel). Besides forgetting the technology, important stakeholders were forgotten as well, for example, the auxiliaries preparing the samples, and the cytodiagnosticians, and the cytologists assessing the samples through the microscope.

Technology was part of my nursing education in the early 1980s as in learning how to insert intravenous lines. In Ihde's terms this would co-constitute what it is to be a nurse. But it was as a PhD student that, for the first time, I "saw" technology as a question in itself. But to include technology and the laboratory, which is necessary to understand "early detection" by means of screening, in my thesis, I had to step outside the margins of what was considered an appropriate topic for nursing research (Forss, 2012). During my PhD I read literature on technology from science and technology studies (STS), Actor-Network-Theory (ANT) and sociology and social anthropology. That is, I read "theories" that equate humans and things and that reduce technology to "the social." Today, the challenge in nursing, as I see it, is to find ways to understand human-technology relations that do not absorb the one into the other and that neither departs from utopian nor dystopian views on human technology relations, or positively, to account for technology as an "actor" rather than neutral and to find "concepts" that enable analyses of concrete cases of human-technology relations.

The debate concerning technology in nursing emerged in the mid-1980s (see, for example, Gadow 1984). Questions were raised on whether technology and nursing are incompatible or not, and on whether more technology necessarily means less nursing (e.g. Bernard & Sandelowski 2001, Almerud 2007). The discussion has problematized whether, instead of a nurse-patient relationship, technology "objectifies" patients and imposes itself as a "nurse-technology" relation to the extent that the patient disappears, vanishes. The nursing-technology relation is thus an important area of research, although focus has primarily been on the nursing-patient relations. Not the least in Sweden where many studies have focused on nursing-patient relations, showing its importance for how patients and relatives experience their situation (Willman, Stoltz & Bahtsevani 2011).

What the lamp teaches us is that nursing practice has always involved technology to enhance human sense, alleviate bodily functions and visualizing invisible bodily processes. Besides oil lamps, nursing practice has always involved materiality, for example, bed pans, bandages, intravenous lines, heating pads, stethoscopes, thermometers and bladder scanners. Typically, there is a third, albeit material, actor involved in the nurse-patient relationship and it is thus important to re-conceptualize the dyadic nurse-patient relation into a triadic one; the nurse-technology-patient relation. Technology thus mediates and contributes to decision making, which means technology needs to be understood as an "actor" instead of a bundle of mute, neutral objects (Verbeek, 2005). According to Verbeek (2006) it is important to ask whether this role had been explicitly delegated or not, and he talks about a distribution of responsibility between users (in this case nurses, patients) and designers.

The following way of thinking about the human-technology relation, based on Ihde's work, conceives the relation between the nurse and the patient as a triad and takes into account the necessity that nursing must focus on both the patient and on technology.

First, safe, ethical and person-centered care requires the ability to oscillate between the patient and the technology, as foreground and background. On one hand, good nursing means focusing on the *patient* to ensure patient participation and to support health where technology remains in the background. On the other hand, good nursing also means focusing on *technology*. This is necessary to ensure patient safety and to prevent harm. Focusing on the technology would mean keeping technology in the foreground and the patient in the background. That is, not entirely forgotten but, for a moment, put the patient in background (see Figure 11.1).

Here, I use this photo for as an example of human technology relations, as it involves many aspects of the nurse-technology-patient relation. The nurse's primary focus is the clock (technology in the foreground), measuring the pulse, while the lamp mediates vision—the patient is momentarily in the background. The patient's primary focus is eventually technology as well and the nurse is in the background. That is, although the patient is breathing through the oxygen mask, she is probably constantly aware of the mask (the mask being both an "object" and "medium for" breathing, and thereby only partially embodied). Also, since she is in an upright position, her closed eyes mean focus on the technology rather than sleep.

Second is the patient-technology environment. Today, emergency care constitutes a technology intensive environment filled with sounds and light. Nursing scholars have shown that patients, despite being observed and "seen" by an abundance of technology, may feel invisible (Almerud, 2007). There is an important message there, in which the nurse must intervene.

Third, we have health itself. Health can also be understood in relation to technology. Many countries have prevention programs to detect ill-health at an early stage, for example through Pap smears, mammography and cholesterol test. This means health is not silent, but rather something to confirm by visualizing invisible bodily processes. Technology can eventually foster *health* if health is seen as increased autonomy, for example through hip replacements or insulin pumps.

Last, but not least we have the grandest of all concepts; *the person*. Organ transplantation, inplants and prostheses mean new hybrids between humans and technology. Technology is reshaping what it is to be a person

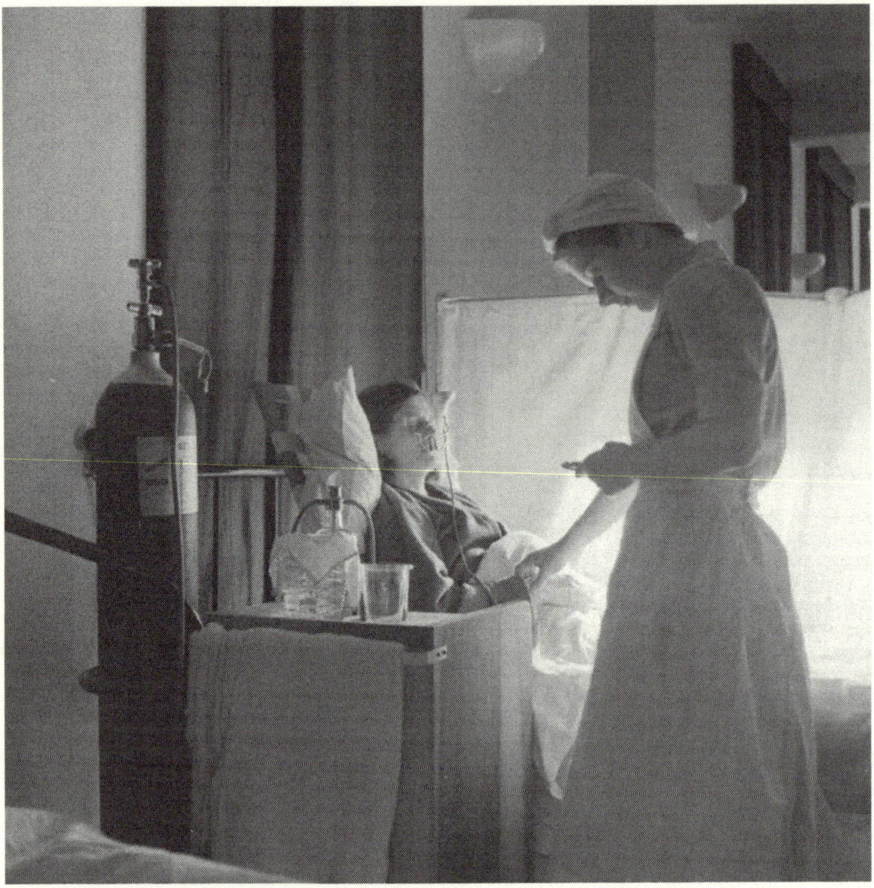

Figure 11.1 Photo released by Imperial War Museum and available for reuse as IMW commercial licence (Catalogue number D 13877). The original photo is entitled "Student Nurse-Life at St Helier Hospital, Carshalton, Surrey, 1943." The original explanatory text is: "Student nurse Joyce Collier takes the pulse of a female patient who is receiving oxygen at St Helier Hospital. Behind them can be seen one of the night screens which are used to prevent disturbance to the rest of the ward." Photographer: Richard Stone. *Source*: © IWM (D 13877)

and is involved in how and when life ends and begins. In nursing, patient-centeredness has begun to move forward in relation to the concept of the Cyborg (Lapum et al. 2012). This latter has a future as yet undetermined. We need to embrace that future yet remains steady at the helm.

I would like to end this essay with a question to the reader who still thinks that human technology relations and technological mediation is outside the realm of nursing. What is the similarity between the following three examples from nurse education, research and practice respectively: self-correcting examinations in web-based education, determining power in sample size in large scale quantitative research, and bed rails to prevent the patient from falling from the bed? It is my belief that nursing philosophy and practice has much to offer the philosophy of technology, to engage with and contribute to it. The lamp is also within.

ACKNOWLEDGMENT

I am grateful to the late John S. Drummond who edited this chapter.

REFERENCES

Almerud, S., Alapack, R., Fridlund, B., & M. Ekebergh (2007). Of vigilance and invisibility—being a patient in technologically intense environments. *Nursing in critical care,* 12(3), 151–158.

Barnard, A. & M. Sandelowski (2001). Technology and humane nursing care: (ir) reconcilable or invented difference? JAN 34(3), 367–375.

Birkler, J. (2007). *Filosofi och omvårdnad.* [Philosophy and nursing]. Stockholm: Liber.

Forss, A. (2012). Cells and the Imaginary Patient: the multistable practitioner-technology-cell interface in the clinical cytology laboratory. *Medicine, Health Care and Philosophy.* 15(3), 295–308.

Gadow, S. (1984). Touch and Technology: two paradigms of patient care. *Journal of religion and health.* 23(1), 63–69.

Ihde, D. (1990). *Technology and the lifeworld—from garden to earth.* Bloomington, Ind.: Indiana Univ. Press.

Ihde, D. (1991) *Instrumental Realism. The interface between Philosophy of Science and Philosophy of Technology.* Bloomington, Ind.: Indiana Univ. Press.

Ihde, D. (2011). Husserl's Galileo needed a Telescope! *Philos. Technol.* 24: 69–82.

Lapum, J., Fredericks, S., Beanlands, H., McCay, E., Schwind, J., and J. Romaniuk (2012). A cyborg ontology in health care: traversing into the liminal space between technology and person-centred practice. *Nursing Philosophy,* 13(4), 276–288.

Latour, B. (1993). *We have never been modern.* Cambridge, Mass.: Harvard Univ. Press.

Tishelman, C., Forss, A., Sachs, L., Lundgren, E.L., Widmark, C. and S. Törnberg (1999). Research on risk and risk in research: theoretical and practical experiences from a multi-disciplinary study on cervical cancer screening in urban Sweden, *Qualitative Health Research.* 9, 45–60.

Verbeek, P.P. (2005). *What things do: philosophical reflections on technology, agency and design.* University Park, Pa.: Pennsylvania State University Press.

Verbeek, P.P. (2006). Materializing Morality: Design Ethics and Technological Mediation. *Science, Technology & Human Values 31*, 361–380.

Willman, A., Stoltz, P. & C. Bahtsevani (2011). Evidensbaserad omvårdnad: en bro mellan forskning och klinisk verksamhet [Evidence based nursing: a bridge between research and clinical practice]. Lund : Studentlitteratur.

Chapter 12

Multistable Roboethics

Cathrine Hasse

INTRODUCTION

In many descriptions of robotic designs within the field of Human-Robot Interaction (HRI) robots are depicted as quasi-biological creatures with potential agency resembling that of human beings. The descriptions seem to fit with a "singularist" perspective envisioning that within only a few decades human intelligence will be replaced by machine intelligence (e.g. Kurzweil 2005, Moravec 1988, 1999). This perspective assumes that technologies, like robots, are stable artifacts that soon can act independently of the humans. It also assumes that humans are exceptional and that their exceptionalism is their intelligence. This exceptional quality may be transferred to and enhanced in machines—beyond our human cultural-material imaginaries. A human ethics would in this perspective entail that advanced robotics does not pose a threat to mankind but will only transform and enhance us for the better.

In the postphenomenological perspectives discussed by Don Ihde and among others Peter-Paul Verbeek and Robert Rosenberger, ethics must be seen as distributed. This distribution is, I shall argue, tied to what counts as ethics in local settings understood as cultural ecologies (Hasse 2015). Ethics emerge in amalgamations of robots and humans in complex cultural realities, where they create somewhat unexpected trajectories for each other. In this chapter I shall argue that roboethics need to take account of these complex realities in which a robot, no more than a human, can be said to possess an inherent stability let alone an agency or an ethics entirely its own. Human exceptionalism is not an overarching intelligence but rather an overarching fantasy about the creation of decontextualized artificial intelligence. Taking "multistability" and "trajectories" into account makes a singularist fantasy of a predictable and controlled posthuman future untenable, indeed unethical, and leaves room for a new postphenomenlogical posthumanism, with a "spinozist" way of envisioning roboethics.

A VIGNETTE

In the robotics literature creatures, with cartoon-like, human-like features built to engage with humans, are often called "sociable robots" (Breazeal 2002). They look "cute" and have hybrid-human-animal features and reactions. These robots are still in the testing-phase; yet in Denmark they can be experienced as technology-in-use in a number of places.

Let me, in the spirit of Don Ihde's own style of including personal experiences in relation to discussions of the widest aspects of entanglements of machines, cultures and embodied perceptions, begin with a small tableau from fieldwork at a nursing home in Denmark.[1] We are sitting in the sun outside one of the houses at a wooden table. I have put a tape-recorder on the table in front of me, the center leader pours coffee and center staff help themselves to buns. Meanwhile we are talking about how terrible it is gradually to lose your mind. One of the staff helpers recounts how the disease Alzheimer's gradually transforms old people in the nursing home from relatively communicative to the mere presence of a human form with almost no contact with the surrounding world. Often the old people just sit in chairs like empty shells with no signs of being alive. Yet, she points to two other creatures joining our coffee table, these two relative newcomers to the staff may still elicit reactions from the old people. The two creatures are placed on the table in a straw basket like the one used for dogs. They are completely alike and completely still. Out of their mouths sticks what appears to be pacifiers tied to an electrical cord. After a while, the center leader declares that the creatures are "ready" and all of a sudden one of the creatures, in the shape of a lilywhite baby harp seal, turns its cute head towards me and says a little appealing sound. It flaps its eyelids and looks at me with huge black eyes and contact is made. Its name is Paro, it's a sociable robot and I immediately get a sense of having a personal contact with what I know is a machine made of nuts, bolts, sensor systems, algorithms and electrical currents. Though the two Paro robots seem identical, I feel a special attachment to this one because it picked me to look at first. My pre-sentential contact does not include thought about how Paro affects the old people or the staff. I am simply charmed by the creature.

Later, however, the staff helper recounts how some old people reject Paro, while others become almost too much attached to the robots. So much so that they in fact will not let it go when Paro is to be passed on to another citizen at the nursing home.

Our relations to robots are, following Don Ihde, alterity relations where the technology is engaging us as a quasi-other (Ihde 1990, 98–100). It is not the technical properties of robot technologies that become focal in this engagement. Inspired by Emmanuel Levinas this kind of human-technology engagement can in a modified sense be perceived as "an ethical, face-to-face encounter" (Ihde 1990, 98). As a forerunner of much later discussions of technology and ethics (e.g. Verbeek 2011) Ihde acknowledges that humans relate pre-sententially to technologies as "focal entities that may receive the multiple attentions humans give the different forms of the other" (Ihde 1990, 107). A number of ethical issues arise when Paro becomes technology-in-use receiving attention as quasi-other in this particular nursing home. It is for instance a problem when a robot runs out of electricity (fed to it by the "pacifier") and suddenly lie still "as if it died," as the helper explains. That makes the old people frightened and sad, the center leader Hanne tells me.

Besides Paro, other robots are to be found engaging with human beings in rehabilitation centers in Denmark. One of them is a round, penguin-like creature on wheels, named Silbot. Silbot is a machine that looks like a cartoon becoming alive. It speaks several human languages and can flap mechanical wings when it gets excited or angry. It is presented for the citizens (with mild forms of dementia) as a "brain-training teacher." Formerly, back in South Korea, it was developed as an English teacher, but here in Denmark it has found new use as a brain-trainer. We met in a training facility where it wheeled around correcting the brain-training exercises made by citizens, examining their answers and encouraging the citizens to continue their training. One day a person, training with Silbot, ran into it and tripped over one of its wheels and exclaimed: "Sorry, I kicked you."[2] When engaging with these robots humans seem perfectly willing to address them as if they were living creatures understanding words. Both staff and the citizens engaging with the robots are perfectly aware the robots are not humans—yet they are not mere machines either. They are the perfect example of Ihde's "quasi-others" demanding ethical responses.

MULTISTABILITY

In STS (Science and Technology Studies) the argument: "guns don't kill people—people kill people" has been fiercely debated. Both ANT (Actor-Network theory) and postphenomenological approaches agree that technology is not a neutral instrument depending on human decision making. Neither is it substance acting in its own right (Verbeek 2005, 136). Following a postphenomenological perspective technology is embedded in lifeworlds of inter-engaging humans and technologies (Ihde 1990) and any ethics has to consider that neither human nor technology act separately from each other but create each other through processes of "multistability" (Rosenberger 2014). This point is becoming increasingly important in a globalized world, where technologies travel freely across borders and cultures (Hasse 2013). Stability is not embedded in the "thing" but in the material as well as traditions and relations following embeddedness in cultural use: "The very structure of technologies is multistable, with respect to uses, to cultural embeddedness, and to politics as well. Multistability is not the same as neutrality. Within multistability there lie trajectories, not just any trajectory, but partially determined trajectories" (Ihde 2002, 106). Considering multistablity, metal boxes created for preserving food in cans in a Western context may become valuable ornaments in another cultural context (Ihde 1990, 125).

Though we may often be unaware of our multistable material hermeneutics, multistability may also lead to controversy and disagreement that can follow from multistable readings of for example, scientific imagery (Rosenberger 2011). This implies that we, in our analysis of roboethics, must go beyond what is directly presented to the eye and explore multistability as a complex process striving towards stability—including the ethical demands and cultural imaginaries that partake in creating stability in the meaningful use of sociable robots as technology-in-use.

In *Chasing Technoscience* (Ihde and Selinger 2003) Ihde discusses the problems with the relativistic approach in relation to Bruno Latour's discussion of the "human-gun" problem. Where Ihde's relativism concerns the inter-actional becoming of humans and guns in cultural contexts, Latour dismantles the notion of culture as well as context and is left with descriptions of human and non-human actors. Ihde (1990) focuses on the kinds of relations which take the human body as the point of departure (embodiment, hermeneutic, alterity and background). Latour is not interested in hermeneutics, which is after all a human invention. Latour focuses on how technologies make material mediations of the actions of human as well as non-human actors for example, the speed bump that makes cars slow down. This is a kind of material ethics where ethics is delegated to the technology (see also Verbeek 2011). Technologies not only reduce or amplify how humans act in the world—technologies make an active contribution to human decision-making (Verbeek 2006a).

Whereas Latour is satisfied with descriptions, Peter-Paul Verbeek moves the discussion into a normative realm taking up a discussion of how technologies come with scripts: "When technologies coshape human actions, they give material answers to the ethical question of how to act. This implies that engineers are doing 'ethics by other means': they materialize morality" (Verbeek 2006, 361). This argument places ethical responsibility on the designers; however, it gives no answer to what Ihde has called "the designers fallacy"—the belief that a designer can "design into a technology, its purposes and uses" (Ihde 2008, 51).

I argue that the human/non-human de-cultured approach becomes problematic from an ethical point of view because human-technology relations cannot be separated from the local settings in which ethics emerge as tied to struggles for stability when technology is put to use. The designers' fallacy is in this perspective linked to the multistability of material artifacts.

In the case of sociable robots the question gets a further complication. Humans are apparently willing to ascribe human-like qualities to machines with cute features and more; the new machines are not just "human-like," they are cute in a way, which makes them stand out from other machines and tools as well as humans. Sociable robots like Paro and Silbot are becoming a new category of "robotic being" demanding ethical work done by the humans engaging with them because they are cute. At the same time these new companion species are global creatures, which become ethically reconfigured as technologies-in-use.

There are (at least) two ways of dealing with this development ethically. One way, which I term the "spinozist" way underscores the entanglements of ideas and bodies in cultural settings. However, some robot designers see robots like Paro and Silbot as forerunners of a future, where machines, not humans, embody the most intelligent, predictable and preferable morality. This is the "singularist" fantasy proposed by robot designers.

SINGULARITY

Some robot-designers hope for a posthuman "singularist" future built on an "exponential growth" in machine intelligence proposed by Kurzweil (2005). The prediction of "a singularity" was first formulated in 1993 by the science fiction writer Vernor Vinge as a special invitee at a NASA conference called "Vision-21":

> Within thirty years, we will have the technological means to create superhuman intelligence. Shortly after, the human era will be ended. [. . .] I think it's fair to call this event a singularity. (Vinge 1993, 11–12)

The transhumanists and the singularists have high hope for the human-machine amalgamation in the future and even dream of transforming humans into streamlined posthuman machines. Natasha Vita-More has for instance in her fantasy and as a piece of art created the perfect hybrid of a human-robot, a "primo-posthuman" which has "smart skin" and a "meta-brain" where super-intelligence can be uploaded (More and Vita-More 2013, 76).

This "singularist" fantasy sees these new amalgamations as more intelligent, and thus better at making moral judgments than humans. Any progress towards the singularity is preferable.

Science fiction has always played a part in the minds of scientists—and engineers and roboticists are no exception (Hasse 2008, 2014). Techno-fantasies such as these may begin at a young age—and already school children have many ideas about robots, what they look like and what they may do (Ihde 2002, xii). According to Cynthia Breazeal she drew a good deal on inspiration from the movie series Star Wars when she constructed her first robots at M.I.T. Media lab in the 1990s. She says: "What kind of future do you envision with robots? The science fiction stories and movies that fascinated me as a child shape my dream. There are the mechanical droids R2-D2 and C-3PO from the movie Star Wars. There are many wonderful examples in the short stories of Isaac Asimov, such as Robbie" (Breazeal 2006, 79). Even though many roboticists, including Breazeal herself, underline that actually creating such robots in 3D real-life versions may be way ahead in the future, others insist that this future is only a few years away—and others that it is already here. When we begin looking for sociable robots with real-life bodies in public media—movies, Youtube videos, and pictures—they often look as well-functioning as those in Star Wars. They move and act as stand-alone entities and we may look into their often very cute eyes and detect the companion souls and consciousness we long for. Professor Junichi Takeno at Meiji University in Japan for one claims that he has created a "conscious robot"—because it has been able to recognize itself in the mirror-test (Takeno et al. 2005). Yet meeting such robots in real life is disappointing. When the roboticists use phrases like "cognition," what they refer to is machine-"cognition"—a machine decoding and reacting on translations of images into points and algorithms.

From that point of view it makes perfect sense to begin wondering about what a roboethics could look like. Following Verbeek many have already tried to think of ways to make engineers morally responsible for the machines they build. "Roboethics" is the abbreviation of "robot ethics" (Bekey 2012, 17). The first substantial contribution to this emerging field of concern emerged in a European network named EURON (European Robotics Research Network) with the task of "mapping" robot ethics. Veruggio, inventor of the phrase "roboethics" in the EURON network, makes clear that "the roadmap does not address the robot and its artificial ethics, but the human ethics of the robots' designers, manufacturers and users" (Veruggio 2007, 7). Ethics is defined by Veruggio as the study and evaluation of human conduct and thus concerned with our relationship to others, whereas morality is concerned with consciousness and our relationship with ourselves. Keith Abney, a philosopher of science ethics, enhances this definition to include a focus on robot conduct when he adds: “First, it can refer to the professional ethics of roboticists (often termed ‘roboethics’ [Veruggio 2007]); second, it can refer to a moral code programmed into the robots themselves—the moral code the robots, not the roboticists, follow; and third (the possibly nonexistent meaning), ‘robot ethics’ could refer to the self-conscious ability to do ethical reasoning by robots—to a robot’s own, self-chosen moral code” (Abney 2012, 35). Abney rightly underlines that the last aspect of robot ethics may never be a relevant approach even though many roboticists and philosophers take as the self-evident basis of their arguments that robots one day will become autonomous “others”—the closest to a meeting with an intelligent alien human beings are likely to meet in a foreseeable future.

ROBOTIC LAWS

On the road to singularity the ethical laws guiding the process of creating super-intelligent machines are often referred to as the three laws of robotics originally formulated by science fiction writer Isaac Asimov—and today widely accepted as a challenging guideline for roboticists (e.g. Brooks 2002, 172–175).

They are most clearly spelled out in the wonderful story “Runaround” from 1942, where we encounter the normally very capable robot Speedy in the deadly sunlight of the planet Mercury, year 2015, running frantically around a Selenium pool, while performing an opera song by Gilbert and Sullivan from HMS Pinafore (For I’m Little Buttercup, sweet Little Buttercup) followed by a: “Whoops!” Looking at Speedy from below ground we find two US employees sent out to repair an old mining station. Their problem is that they have run out of the selenium powering their photo-cell banks keeping the temperature at a reasonable level. Without new selenium they will die. They have ordered Speedy to go out, for them, in the deadly sunshine and get it from the pool. However, instead of obeying orders as it is supposed to, now Speedy is running around like it has been drinking. It turns out that because of a weak order, formulated by one of the men, an apparently unsolvable double-bind crisis has evolved in Speedy relating to two of the three basic roboethic laws:

1. A robot may not injure a human being or, through inaction, allow a human being to come to harm.
2. A robot must obey the orders given to it by human beings, except where such orders would conflict with the First Law.
3. A robot must protect its own existence as long as such protection does not conflict with the First or Second Law.

The problem is that Donovan who gave Speedy the order did not underline strongly enough that the order should be obeyed even if the robot puts itself into danger (because of volcanic activity near the selenium pool):

> "*Donovan:* 'I said . . . uh . . . I said: 'Speedy, we need some selenium. You can get it such-and-such a place. Go get it.' That's all. What more did you want me to say?'
>
> *Powell:* 'You didn't put any urgency into the order, did you?'" (Runaround—Asimov 2004, 27)

Due to the lack of urgency the robot goes in circuits following one moment the law number 3 and the next law number 2 and then 3 again. The crisis is only solved (and the men saved) when after a while Powell make Speedy obey the law number 1 by putting his own life at risk—thus forcing Speedy out of the confused situation.

In the real work of robots and roboticists Asimov's laws are well known and often considered a hard-to-create but appropriate guideline for robot designers (Brooks 2002, 172). This approach to roboethics is, however, what Keith Abney calls "top-down":

> Top-down, rule-based approaches, like Asimov's Three Laws of Robotics (Asimov [1942] 1968), understand ethics as the investigation of right action—what are the rules to follow in order to be morally right, to perform the morally correct. (Abney 2012, 36)

In Asimov's story Speedy is far from the singularists vision of an intelligent robot. Ethics is distributed between the engineers designing Speedy, Speedy and the (for now) more intelligent humans striving to stay alive even if it means sacrificing the robot.

From a postphenomenological perspective we could argue that the top-down rule-based roboethics overlook that artifacts as well as ethics become culturally embedded, when technology is put to use. Robots are multistable and no matter what rules they have been designed to follow ethics is formed in relation to local trajectories.

No entity—no more a robot than a human—can be understood as a stand-alone entity with sole inbuilt responsibility for an action. Concerning ethics we have to include what humans and machines do together in complex entanglements. No single cultural interpretive ethics can be said to be the only relevant one. In pluricultural times we are in need of "compound eyes" (Ihde 1993, 114, Verbeek 2005, 138). In this amalgamation of machines, human beings, ideas of cuteness and soft or hard materials there are, however, stabilizing trajectories. The main advantage over the crude relativism found in much posthuman literature is that with a postphenomenological perspective we may avoid any naïve researcher generated presentism by following processes of learning how matter stabilizes in local cultural ecologies (Hasse 2015).

The mediation of artifacts is not between a separate subject meeting a reified object but the mediation rather forms a mutual constitution of both subject and object (Verbeek 2005, 130). The constitution of trajectories can be said to follow learning paths working on obtaining stability in relation to an already existing practice of how subject and material form tools and signs in cultural practices (Hasse 2013).

Thus when our concern is the alterity relation, and the ethical consideration it brings about, sociable robots are neither a stand-alone subject nor a stand-alone object. The constitution of the alterity relation must take the whole cultural practice into account. Thus (contrary to Bruno Latour's position) we need an ethic which takes the cultural complexity of trajectories of constantly (re)emerging assemblages of robots humans in their otherwise formed environments into account. Though Ihde largely reserves his discussions of how perception is amplified and simultaneously reduced by technology to his discussions of embodied and hermeneutic relations, I shall argue that alterity relations also are affecting our embodied material hermeneutics when sociable robots show themselves to be multistable as they "jump" from their smooth screen or designer existence into people's everyday lives in hospital wards, nursing homes and schools.

Here any singularist illusion is quickly put to the test. Instead of meeting the robots of roboticists science fiction fantasies—we met a new companion species stretching our ethical work to include quasi-others—in all their faulty existences. The inmates of nursing homes find themselves confronted with how to deal with these often very badly constructed but "cute" machines. This can be quite a shock for instance to discover that "cute" robots are very noisy thus breaking any illusion of an intelligent being. One staff member interviewed noted about Silbot:

> The noise . . . it is terrible because there is an engine noise. Maybe you can also hear it a little in the background? [Our users] are initially very annoyed [with the sound]. Eventually when you have been here for a week or so, we do not hear the noise anymore because we are concentrating so much on getting the job right. (Centre staff, Louise)

In Ihde's terminology the noise has become a "background" relation leaving room for an alterity relation (1990). What is often overlooked, even in postphenomenological analysis, is the hard work needed to stabilize robots as proper ethical alterity relations.

Multistability "highlights the fact that a particular technology cannot be forced into just any context, or used for just any purpose; due to the technology's concrete materiality, only some uses prove experientially coherent or 'stable.'" (Rosenberger 2010, 135). Yet people prove willing to put up with a lot of trouble, including noise, in order to be able to engage with robots like Silbot. And it's not because it's emerging as especially ethical or even pleasant.

The robots implemented at Danish nursing homes and rehabilitation centres are clearly not very intelligent machines. Yet we are willing to grant them "liveliness" and are caring for them when they stop functioning. It becomes a concern for the citizens as well as staff, when Paro goes dead or Silbot is stepped on.

It seems as if gradually humans adjust their way of engaging with the machines as quasi-others. Stability is hard work and it seems to be worth the labor not just because of the actual machines, but because of the fantasies tied to them—which again are more about social expectations than intelligence inside the machinery.

A SPINOZIST PERSPECTIVE

If we accept the argument of the designers fallacy as tied to the argument of multistability there is still work to be done creating a posthuman ethics including a postphenomenological perspective. This perspective could be called "spinozist" and contrasted with the perspective found around the "singularist" versions of a posthuman future. It is "spinozist" in the sense that a sociable robot as well as a human being could include their respective striving for existence—conatus (Spinoza 2004)—yet for a while be seen as performing multistability in complex assemblages.

In the spinozist ethics all things actively strive for existence, yet no individual body can be considered independent of other bodies. All bodies affect each other in their desire to fulfill their potentialities. It is when conatus is aligned that Spinoza notes that "if two come together and unite their strength, they have jointly more power, and consequently more right over nature than both of them separately, and the more there are that have joined in alliance, the more right they all collectively possess" (Spinoza, 2004, 296).

Though the subject-object distinction is dissolved, humans and non-humans are bodies—not as a substance that, in spinozist lingo, is the cause of itself and separate from other things, but as bodies affecting each other. Bodies are as "interveners" more forceful than material entities (Bennett 2010). Yet they meet in particular cultural ecologies forming their ethical encounters.

In these undetermined movements of mass and masses we humans cannot take control of what each body is doing to affect the mass of bodies. The claim that engineers should take moral responsibility for the robots they built may be a sensible demand—yet roboethics must include more than engineers and their products. In amalgamations where bodies "are finite modes of being that need to be acted on and are dependent on other beings" (Kagawa and Moro 2009, 183) one can have a moral responsibility for the ethics—not of another body but of the amalgamation of bodies one engages with. In these assemblages humans are decentered as "controllers"—and thus posthumanism gets a new meaning that differs from that of the "singularists." In this sense of "decentered humans" we have always been posthuman—as proposed by Kathrine Hayles (1999). With the new spinozists interpretation Jane Bennett argues that:

> Bodies enhance their power in or as a heterogeneous assemblage. What this suggests for the concept of agency is that the efficacy or affectivity to which this term has traditionally referred becomes distributed across an ontologically heterogeneous field, rather than being a capacity localized in a human body. (Bennett 2010, 23)

In the trajectories of supporting, conflating or confronting other bodies striving for their existence the decentered human is constantly in a process of becoming. The notion of the autonomous stable human is a multistable posthuman in processes of cultural-historical changes of all kind of stabilities (thus avoiding the predominant "presentism" informing much postmodern theory). In this line of thinking robots like Paro and Silbot and humans are nothing in themselves—but their stabilities are formed in cultural amalgamations.

Stability emerges at quotidian encounters for example at the coffee table in my meeting with Paro. The assemblages are not reduced to the present bodies of staff, centre leader and robots, but include the leader of the nursing homes contact with the director of technological innovation at the local municipality, functioning electrical currents, manga cartoons, a politicians accidental meeting with a Japanese robot engineer, and my long-time relationship with a cute cat at home (Hasse 2013). My perception of a "soul" and "recognition" in Paro's black eyes under the long eyelashes, experiencing a real contact, may make me want to help preserve this cute creature. I know it as belonging to a world of fantasy where the good always wins and the good is white and innocent and seemingly helpless as the robotic harp-seal, which must be protected by me. Yet in another amalgamation I may want to silence it forever—putting it on the attic as a used toy to be forgotten. Like the staff make me aware (also found in a study on Paro by Sherry Turkle and her colleagues, Taggart et al. 2005) some citizens come to hate the creature and do not find it cute at all (Hasse 2013). Not everyone in our culture loves cartoons—yet they are increasingly confronted with 3D versions as a new kind of being they have to learn to deal with. Robots emerging in such amalgamations are anything but dead substance—they are vibrant materiality (Bennett 2010) which transforms who we are and thus we need to discuss what a posthuman ethics could look like.

In meeting Silbot at the rehabilitation center we rather seem to be in the age of 'ante-robocreatures' than 'primo-posthuman.' Silbot is not trying to pass as a human, but is clearly a machine-like creature. And for a while it is a very multistable one indeed. Developed in South Korea as an English teacher and brought to Denmark as a brain-trainer, it's stability is destabilized—yet it emerges as a new category of being that people will begin to treat as a body with a legitimate presence and that they will have to learn to engage with and reinforce.

> There were actually several who said they thought SILBOT was not important. Then I confronted them and asked them: "Well, you said SILBOT was unimportant. So why did you then walk over and said 'have a nice weekend' to it?" (Line, Centre staff)

Now staff and citizens treat Silbot as they would each other—greeting it politely. Before it came to this cordial relationship Silbot had to be reconfigured in order to take part in the amalgamations formed at the rehabilitation center. The problem was that the robots' brain-training program developed in South Korea was directly translated into Danish. This translation turned out not to fit the cultural context of the Danish rehabilitation center and its citizens. In the direct translation into Danish the "teacher" seemed to speak perfectly understandable Danish. However, when the robot was put to use at the nursing home it became apparent this in the direct translation from Korean into Danish made Silbot into a rude bully in need of a lecture or two on politeness, due to the cultural differences of addressing people in the two languages. When it was put to use in public, it turned out the citizens perceived the robot as scolding them with rough words for not getting the answers right in their brain-training exercises. Robots as artifacts are not *carriers* of culture. It was in the meeting with the local cultural ecology, that the nursing homes staff expectations of how a robot teacher should, or should not, address citizens emerged. Here Silbot turned out to be conceived as a rude and demeaning cultural creature that had to be stabilized through re-programming.

> It's been reprogrammed after it has come to Denmark. It is not as angry, hard and cold anymore as when it came. In Korea you have a winner and a loser. So it's a completely different culture. It has been programmed in a different way because it simply scolded the participants when they answered incorrectly. It had a completely different cultural approach to learning than we use in Denmark. (Erica, centre staff)

The sociable robots turned out to make many unexpected ethical demands on the citizens and the staff in order to ensure robots had the "right" ethical behavior. The work done by Paro and Silbot could easily be done by humans. Paro and Silbot have not been designed to fulfill a particular need as technology-in-use demanded by the local staff. There is no real need for Paro as a soother and Silbot as a brain-trainer. They have been designed to be quasi-others engaging with humans on the road to a fantasy future, where cute sociable robots will engage meaningfully with humans as a new companion species. The staff and citizens have to do a lot of hard work to include these bodies in their local amalgamation. Even when "corrected" the staff and citizens have to keep learning how to stabilize this new category of being. Even so the presences of the robots are never questioned. In the process, robot, staff and citizens gradually became stabilised bodies in an amalgamation including material bodies as well as ideas of a robotic future. The main bulk of stabilizing work however lay on the staff and citizens.

Morana Alac has together with a group of researchers from San Diego explored what give robots sociality in robotics laboratories. Is it, as it appears from the seductive advertisements presenting the robots in public media, an inbuilt function of the robot? Alac and her colleagues come up with another answer.

Though social roboticists have designed their robots to function as social agents in interaction with humans and other robots Alac et al. concludes that the "spatial organization and coordination between the robot and the humans who interact with it" is of utmost importance (Alac et al. 2011, 893).

Alac et al. is studying a so-called RUBI-robot in its interaction with robot designers and toddlers between 18 and 24 months of age and they are inspired in their work by visits to a preschool. RUBI is a robot designed to help toddlers learn, for example, simple geometry, but just like in the Danish cases it seem that the robots are not designed to fulfill an already established need (of, for example, nursing or teaching better than any human). The real purpose of developing robots does not seem to fulfill a particular purpose of being sociable, but seems to aim at experimenting with how far you can stretch human engagements in making the robot function in human-machine interactions.

Alac et al. (2011) argue convincingly that it is the reactions and interplay between the robot designers, social robots and other material artifacts and small pre-school children that make the robot come across as social beings—just as much as it is an inbuilt quality in the robots themselves. They do not mention how robot cartoons have already been part of the human's social environment but also RUBI has clear cartoon like features just like Silbot. It is a quasi-other in its own right—a new companion species we strive to stabilize in our local cultural ecologies.

It is the combined "cuteness" in the material and stabilizing social interaction work that make it possible to give the robot a presence as a being—more than any inbuilt quality of the robot. In fact the noises of Silbot disturbing its cuteness become a background relation as a result of this interactional work partly responsible for holding the amalgamation of bodies together.

> Here we suggest that to deal with this issue one must look well beyond the robot's appearance and built-in control and cognitive architecture. However, we do not mean to consider only the discursive strategies that present the robot as human-like or to reduce the problem to human predispositions to anthropomorphize the robot. Moreover, we do not believe that an account of a robot's interactions with a single individual would fully resolve the problem. Instead, we claim that a subtle interactional coordination between multiple human actors—including the robot's designers—is crucially important for sustaining human interest in a robot. We further claim that this multiparty interactional coordination allows a technological object to take on social attributes typically reserved for humans. (Alac et al. 2011, 894)

HUMAN STRETCHING

The reason why I quote Morana Alac and her colleagues at length is because I see a tendency in posthuman theorizing to overlook the importance of human perception of, for example, cuteness in creating "cuteness," and the importance of agential sociable humans in their embodied and sensuous and fantasy driven relations with technologies in new posthuman theories.

We must take ethical responsibility for how these fantasies of robots as companion species materialize and put demands on human "stretching" of themselves as well as the robots. When the staff decided to reprogram Silbot they made an active contribution to how the technology as quasi-other can be stabilized as cute. Humans stretched the technology so Silbot now contributes to teach human newcomers how to address each other politely.

Are humans more ethically responsive and ethically responsible than the machines we encounter? I find an asymmetry when dealing with actual robots as technology-in-use, which could be further explored. Humans, more than robots, demand and give ethical responses. Human, more than robots, recognize hard and soft materials and electrical currents as "cute." Citizens say: "I am sorry" if they step on the robots toe, and they also demand that the robot is re-programmed so it can come up with ethically acceptable responses to the citizens. Thus the ethical stability of robots first makes demands on human agency and then on robotic agency. Even though "technologies contribute actively to how humans do ethics" (Verbeek 2011, 5) it is after all *humans* who demand ethical responses.

It is not just of concern for roboticists—but a concern for all of us when engaging with sociable robots without questioning our own human fantasies included in amalgamations.

As noted by Ihde our fantasies have so far moved from organic to technological:

> From flying carpets to warp speed, I note there is little new in such fantasies. The difference is that, since modernity, the fantasy embodiments have tended to be technological rather than organic, animal-like, or supernatural. (Ihde 2011, 125)

Yet now they also seem to move from cartoon fantasies to materialized technology. Robots like Paro and Silbot have also moved from being fantasies materialized in words and writings to become materialized as 3D entities changing not only humans but also our material amalgamations. It seems as if we strive to stabilize these as our new companion species at all costs, no matter how much we have to stretch them and us in order to include them in our amalgamations.

Robert Rosenberger has discussed how the notion of multistability can be used for critical analyses that are destructive or constructive. Whereas constructive analyses "explore how conventional understandings of technological relations can be productively reconceptualised as multistable, prompting concrete investigations of the details of particular stabilities" a destructive analysis is "dismantling and exposing overly optimistic accounts" (Rosenberger 2010, 135). This is an important point. In that sense my use of multistability may here be more of the destructive kind than the productive kind. Yet, maybe we need a destructive analysis first in order to become productive. When Ihde praises the technology that relieves humans from doing hard work he is not taking into account that technology also demands hard work—work that changes who we are and how we relate to each other (Hasse 2013). A productive postphenomenological analysis must explore new possibilities for sociable robots from an analysis which includes both implications on how sociable robots relieve us of hard work—but also demand the hard work of stretching humans and robots in order to reach undisputed fantasies.

In the singularist perspective roboethics is giving preference to the immortality of machines over the mortality of deadly humans. The robot is not discussed as a quasi-other relation but as a replacement ethics where human parts are replaced and enhanced by better machine parts. This changes the relation from one of quasi-other to an embodiment relation (Ihde 1990,72). In the spinozist perspective on roboethics there is an attention to how humans and robots together create well-being in local amalgamations, where both technology and humans contribute actively to ethical stabilizations. No doubt face-to-face encounters with Paro may help citizens with Alzheimer endure life. However it is also hard work for the staff, when they have to remove a companion species from one old person to pass it on to the next. They must find ways to stabilize Paro in ways which keep its comforting effects while not creating new uncertainties.

An ethical asymmetry also seems to appear in the case study on Silbot. What is ethical or unethical in the robot is defined by humans in the local ecologies. Being rude or cute is not built into the robot—it emerges in the particular context. The staff perceive the robot agency as "South Korean behaviour" (whereas other parts of the "cute" technology are not acknowledged as particularly South Korean). This rude behavior challenged the well-being of citizens (though possibly not citizens in South Korea). Furthermore—even though they perceive the quasi-other as unethical they do a lot of work to keep it in their local ecology (Hasse 2015). Humans stretch themselves to remedy Silbot's unethical behavior—just as Paro staff stretch themselves to find ethical solutions to the existence of Paro near their citizens with Alzheimer.

Though humans in an STS-perspective are truly decentered in our own self-understanding, we are still responsible for disturbing the fantasies of the future that co-produce futures we may not want. Thus an ethic is needed that will not unquestioningly make humans and robots partake in amalgamations which strive towards a singularist vision of a future, which manifest a largely unacknowledged "cute" fantasy of a cartoon-world.

TOWARDS A NEW POSTHUMAN ETHICS

In a spinozist perspective the human is, just like in the "singularist" vision, a decentered human. Yet the decentered human in the spinozist posthumanism is not to be replaced by a posthuman super-intelligence. Instead of a top-down ethics we need an ethics which can take account of multistabilities.

This means, following the argument of designer fallacies and multistability, that engineers cannot solely be held entirely morally responsible for having fantasies, and attempting to build (or not build) "intelligent" robots like Speedy, robot-like humans like Primo or cute robots like Paro. Neither can they be solely blamed for coming up with "disappointing" robots like Silbot. Ethical problems cannot be solved just by letting engineers try "to activate and articulate as many actors as possible and make passive as few as possible" (Jensen 2003, 235). No matter how much they stuff into the new companion species, robots will be multistable bodies emerging as more or less ethical quasi-others in local amalgamations.

Multistability indicates that the stabilizing process always takes place in *particular* cultural ecologies like rehabilitation centers or, for example, scientific environments (Rosenberger 2008, 2011). Robots are no more cultural entities than humans—yet cultural responses emerge when they meet. Meeting robots involves changes in humans, but may also involve changes in materiality in order to obtain stability (as in the case of reprogramming Silbot—and also Paro is transformed even though it materially may seem untouched—see Hasse 2015 for a detailed account of how cultural changes come about).

The "singularists" envision a future that is anchored in a narrowly Cartesian understanding of human nature (Hayles 2005). The ethical demands are solved by creating an increasingly autonomous intelligence that can substitute fallible humans. We can be in control of this process because machines follow rules whereas humans are fallible and biologically weak. In a spinozist perspective we are past this conception of a human being. The "spinozist" vision underscores that humans have always been decentered and are emerging in assemblages (e.g. Hayles 1999, Bennett 2010). However, both approaches suffer from a certain "presentism" in relation to a posthuman ethics overlooking multistable processes of becoming. The first reduces human nature to a stable understanding of "intelligence" which does not take multistability into account but makes "intelligence" an absolute stable good to be transferred from humans to be enhanced in machines. The second approach fits with the multistable perspective in here-an-now amalgamations, but overlooks that stability is hard work and obtained through already formed cultural trajectories (Ihde 2011), which leave traces for future generations of amalgamating bodies meeting and engaging in particular emerging cultural ecologies.

Humans have shown that when they engage in amalgamations they are increasingly partaking in radical changes which are not just out-of-control, but also seem to act against the best interest of all human beings regardless of cultural background (Zalasiewicz et al. 2010). They increasingly stretch themselves to allow new bodies, built by their own fantasy, which acknowledge the blurred line between fantasies of cute cartoon creatures and human needs for care. The engineers' fallacy may be that their fantasies are never brought to life exactly as imagined.

A postphenomenological posthuman ethic could be an answer to this predicament. One that both focuses on the process of stabilization in local ecologies and takes a spinozist point of departure—and thus keeps the trajectories of human fantasies present in the amalgamations of bodies.

The new category of cute sociable robots as companion species is not the improved posthuman super-intelligence predicted by the singularists, but a new kind of an alterity-relation in local ecologies, which demands we rethink roboethics from the perspective of multistability.

I find an asymmetry insofar as humans more than robots demand and give ethical responsiveness in order to stabilize robots. Humans seem more malleable and capable of stretching themselves to a greater length than the machines they engage with. Human can thus be held responsible for the kinds of stretched amalgamations where they may reinforce the agency of bodies that is distributed across an ontologically heterogeneous field—including virtual, physical and fantastical space. The roboethics I propose includes considerations of this "stretching" process tied to local cultural ecologies.

NOTES

1. Interviews with the leader, three staff members, observations and talks with people at the old peoples home were conducted by me in Jutland, Denmark 19–20 June 2010 (see also Hasse 2013). All names of informants in the study have been anonymized.

2. The fieldwork at the rehabilitation center, also in Jutland, was conducted by three researchers, Signe Hanghøj, Maja H. Bruun and to a lesser extent myself in 2013. All names of informants in the study have been anonymized.

REFERENCES

Abney, K. (2012). Robotics, Ethical Theory, and Metaethics: A Guide for the Perplexed. In *Robot Ethics: The Ethical and Social Implications of Robotics*. Lin, P., Abney, K., Bekey, G. (eds) Cambridge, Mass.: MIT Press: 35–52

Alac, M. (2009). Moving Android: On Social Robots and Body-in-Interaction. *Social Studies of Science*, 39(4): 491–528.

Alac, M., J. Movellan, and F.Tanaka (2011). When a Robot is Social: Spatial Arrangements and Multimodal Semiotic Engagement in the Practice of Social Robotics. *Social Studies of Science* 41 (6):893–926.

Asimov, I. (2004). [1942]. *I, Robot.* New York: Spectra: 20–33.

Bekey, G.A. (2012). Current Trends in Robotics: Technology and Ethics. In P. Lin, G. Bekey, and K. Abney (eds.) *Robot Ethics: The Ethical and Social Implications of Robotics.* Cambridge, Mass.: MIT Press: 17–34.

Bennett, J. (2010). *Vibrant Matter: A Political Ecology of Things*. Durham: Duke University Press.

Breazeal, C. (2002). *Designing Sociable Robots.* Cambridge, Mass.: MIT Press

Breazeal, C. (2006). Human-Robot Partnerships. In S. Coradeschi (eds) *Human-Inspired Robots, Trends and Controversies*, IEEE Intelligent Systems, July 2006: 79–81.

Brooks, R.A. (2002). *Flesh and Machines: How Robots Will Change Us.* New York: Vintage.

Feenberg, A. (2006). Active and Passive Bodies: Comments on Don Ihde's Bodies in Technology. In *Postphenomenology: A Critical Companion to Ihde*, E. Selinger, ed., Albany: SUNY 2006:189–196.

Hasse, C. (2015). *An Anthropology of learning.* Springer: Dordrecht.

Hasse, C. (2013). Artefacts that talk: Mediating technologies as multistable signs and tools. *Subjectivity,* 6, 79–100.

Hasse, C. (2012). Gender in innovative techno fantasies. In Schilhab, T., Frederik, S., Terrence, D. (eds.) *The symbolic species evolved*, Springer: Hamburg: 263–281.

Hayles, K. (1999). *How we became posthuman; Virtual bodies in cybernetics, literature, and informatics*. Chicago: The University of Chicago Press.

Hayles, N.K. (2005). "Computing the Human." In *Theory, Culture &C Society* 22(1): 131–51.

Ihde, D. (1993). *Postphenomenology: Essays in the Postmodern Context.* Evanston: Northwestern University Press.

Ihde, D. (1990). *Technology and the Lifeworld: From Garden to Earth.* Indiana: Indiana University Press.

Ihde, D. and E. Selinger (eds.). 2003. *Chasing Technoscience: Matrix for Materiality* Bloomington & Indianapolis: Indiana University Press.

Ihde, D. (2008). The Designer Fallacy and Technological Imagination. In Vermass, P.E, Kroes, P., Light, A. & Moore, S.A. (eds.). *Philosophy and Design—From Engineering to Architecture.* Dordrecht: Springer, pp. 51–59.

Ihde, D. (2002). *Bodies in Technology.* Minneapolis: The University of Minnesota Press.

Ihde, D. (2009). *Postphenomenology and Technoscience—The Peking Lectures.* Albany, NY: Suny Press.

Ihde, D. (2011). Of Which Human Are We Post? In G. R. Hansell and W. Grassie (eds.). *Transhumanism and Its Critics.*Philadelphia: Metanexus Institute. Chapter 8: 123–135.

Ihde, D. (2002). *Bodies in Technology,* Minneapolis and London: University of Minnesota Press.

Jensen, C.B. (2003). Latour and Pickering: Post-human perspectives on science, becoming, and normativity, in *Chasing technoscience: Matrix for materiality.* Don Ihde and Evan Selinger, (eds.). Bloomington & Indianapolis: Indiana University Press. 225–240.

Kurzweil, R. (2005). *The Singularity is Near: When Humans Transcend Biology.* New York: Viking.

Kagawa and Moro (2009). Spinozic Reconsiderations of the Concept of Activity: Politico-Affective Process and Discursive Practice. In A. Sannino, H. Daniels & K. D. Gutierrez (eds.), *Learning and Expanding with Activity Theory.* Cambridge: Cambridge University Press: 176–194.

Moravec, Hans P. (1988). *Mind Children: The Future of Robot and Human Intelligence.*Cambridge, MA: Harvard University Press.

Moravec, Hans P. (1999). *Robot: Mere Machine to Transcendent Mind.* New York: Oxford University Press.

More, M. & Vita-More, N. (2013). The Transhumanist Reader: Classical and Contemporary Essays. M. More and N. Vita-More (eds.). New York: Wiley-Blackwell.

Rosenberger, R. (2008). Seeing the World Through Technology and Art. Review of Mediated Vision, P. Kockelkoren (ed.). *Techné: Research in Philosophy and Technology.* 12(1): 90–97, 2008.

Rosenberger, R. (2010). Deflating the overblown accounts of technology: a review of Don Ihde's Ironic Technics, *AI & Soc* (2010) 25:133–136.

Rosenberger, R. (2011). A Case Study in the Applied Philosophy of Imaging: The Synaptic Vesicle Debate, *Science, Technology & Human Values* January 2011 36: 6–32.

Rosenberger, R. (2014). Multistability and the Agency of Mundane Artifacts: From Speed Bumps to Subway Benches. *Human Studies:* (2014) 37:369–392.

Spinoza, B. de (2004). A political treatise. In *A theologico-political treatise and a political treatise* (pp. 278–198). New York: Dover.

Taggart, W. Turkle, S. and C. D. Kidd (2005). An interactive robot in a nursing home. Preliminary remarks. In *Toward Social Mechanisms of Android Science*, Stresa, Italy, July 2005. Cognitive Science Society. pp. 56–61. http://web.media.mit.edu/~coryk/papers/Paro_AndroidScience05.pdf. Retrieved 23.08.2012.

Takeno, J, K. Inaba, and T. Suzuki. (2005). Experiments and examination of mirror image cognition using a small robot, *Proceedings of the 2005 IEEE International Symposium on Computational Intelligence in Robotics and Automation* 493–498.

Turkle, S. (2011). *Alone Together.* New York: Basic Books.

Verbeek, P.P. (2011). *Moralizing Technology: Understanding and Designing the Morality of Things.* Chicago and London: University of Chicago Press.

Verbeek, P. P. (2007). Beyond the Human Eye: Technological Mediation and Post-human Visions. In P. Kockelkoren (ed.), *Mediated Vision.* Rotterdam: Veenman Publishers and ArtEZ Press. 43–53.

Verbeek, P. (2006). The morality of things: A postphenomenological inquiry. In E. Selinger, (ed.). *Postphenomenology: A Critical Companion to Ihde*, Albany: SUNY 2006. pp. 189–196.

Verbeek, P. P. (2005). *What Things Do: Philosophical Reflections on Technology, Agency, and Design.* Trans. R. P. Crease. State College: Penn State University Press.

Vinge, V. (1993). The Coming Technological Singularity: How to Survive in the Post-Human Era. In G. A. Landis (ed.) *Vision-21: Interdisciplinary Science and Engineering in the Era of Cyberspace*, NASA Publication CP-10129/1993:11–22.

Veruggio, G. (2007). The EURON Roboethics Roadmap, European Robotics Research Network, *Atelier on Roboethics*, 2005–2007. <http://www.roboethics.org> (accessed February 20, 2014).

Zalasiewicz, J. Williams, M., Steffen, W. & P. Crutzen. (2010). The New World of the Anthropocene. *Environment Science & Technology* 44 (7): 2228–2231.

Chapter 13

Toward a Theory of Technological Mediation

A Program for Postphenomenological Research

Peter-Paul Verbeek

THEORIZING TECHNOLOGY

Over the past decades, the philosophy of technology has developed into a new subfield of philosophy, with its own domain of inquiry, journals, conference series, educational programs, and forms of societal relevance. But, remarkably, the various approaches that have developed in the field have hardly matured into systematic theories. A typical introduction to philosophy of technology (Dusek 2006, Kaplan 2009), indeed, does not give an overview of the most prominent theories, but rather of the main issues in the field, like technological determinism, technological rationality, and the ethics of technology. Even though there are some good exceptions in subfields of philosophy of technology, like Andrew Feenberg's "critical theory of technology" (2001) and the "dual nature theory of technological artifacts" developed by Peter Kroes and Pieter Vermaas (Kroes and Vermaas 2002), the time has come to develop a broad theory of technology.

At the same time, the empirical field of Science and Technology Studies (STS) has developed from an "empirical detour" to answer philosophical questions regarding science and technology into an established field for the empirical investigation of scientific and technological practices. But in this development, its originary philosophical ambition seems to have moved to the background. STS could benefit from a reconnection with philosophical theory, in order to develop new ways to understand the social roles of science and technology. Besides deconstructing the networks of relations that constitute scientific facts and technological artifacts, which is the focus of much STS research, new empirical approaches are needed to the ways in which technologies help to shape knowledge, morality, etc.

In this chapter I will sketch the outlines of a program to develop such a theory in the basis of the postphenomenological approach philosophy of technology: a theory of "technological mediation" that builds upon Don Ihde's analysis of human-technology relations (Ihde 1990). Postphenomenological studies (Ihde 1993; Selinger 2006; Rosenberger 2008, 2012; Verbeek 2008) typically have at least two things in common. First of all, they study technology in terms of the *relations* between human beings and technological artifacts, focusing on the various ways in which technologies help to shape relations between human beings and the world. They do not approach technologies as merely functional and instrumental objects, but as mediators of human experiences and practices. And second, they *combine philosophical analysis with empirical investigation.* Rather than "applying" philosophical theories to technologies, the postphenomenological approach takes actual technologies and technological developments as a starting point for philosophical analysis. Its philosophy of technology is in a sense a philosophy "from" technology.

In my own work (Verbeek 2005, 2011a, 2011b), I have developed the postphenomenological approach in the directions of agency, ethics and philosophical anthropology, exploring how technologies help to shape moral actions and decisions, and how "human enhancement technologies" like brain implants and tissue engineering urge us to reconceptualize the boundaries between the human and the technological and to rethink the concepts we have to understand the human being. This has resulted in first elements of a theory of technological mediation, conceptualizing various ways in which the boundaries between the human and the technological are fading and how the concept of mediation can help to analyze human-technology relations.

By now, it is time to take the postphenomenological approach one step further. The phenomenon of technological mediation deserves to be studied in a more comprehensive and systematic way, covering the full depth of the various dimensions of the relations between human beings and reality. But in order to arrive at such a theory of technological mediation, one crucial step needs to be made. In all attention for the mediating roles of *technologies*, the processes of appropriating these mediations by *human beings* have remained largely unstudied. In order to develop a full understanding of processes of mediation, we should not only study "what *things* do" (cf. Verbeek 2005) but also how *humans* give meaning to these mediations—both empirically and conceptually. How do scientists actively engage with perceptual technological mediations when interpreting reality? How do moral decisions take shape in the active interplay between material mediations and human appropriations?

In the adjacent field of Science and Technology Studies, processes of technological mediation have been studied as well: several scholars in the field have been working on "material agency," but detailed studies of the phenomenon of technological mediation are typically missing in their analyses. Karen Barad's highly important work on "intra-action," for instance, is closely related to the postphenomenological idea of the "mutual constitution" of "subject" and "object," but it does not explicitly address the mediated character of this intra-action (Barad 2003). In postphenomenological theory, the media of this mutual constitution, and the dynamics of their interactions with human agency and the phenomena being studied, have a central place. Similarly, Andrew Pickering's concepts of the "mangle of practice" and the "dance of agency" focus on the interplay between human and material agency, but only indirectly address processes of mediation. His well-known analysis of the interplay between the actions of scientists and the activities in Glaser's bubble chambers, for instance, could be augmented with a detailed analysis of the ways in which visual presentations of bubble tracks help to constitute the phenomena being studied and the ways in which scientists study them (Pickering 1995).

And while the work of Ian Hacking, to give a third example, explicitly thematizes the roles of scientific instruments like microscopes in the production of scientific knowledge, it mainly focuses on the resulting reality-status of the entities investigated in this way (Hacking 1983). To this highly important work a theory of mediation can add an in-depth study of the technological mediations and appropriation processes involved in knowledge production with the help of scientific instruments. Fourth and last, Bruno Latour's work on technical mediation (Latour 1994) offers a set of concepts to study how programs of action get translated when human beings form associations with technical artifacts. But, as will be elaborated below, his symmetrical approach to humans and nonhumans makes it impossible to investigate the hermeneutic dimension of processes of mediation: the human appropriation of technological mediations around which specific "objectivities" and "subjectivities" crystallize.

In Science and Technology Studies, therefore, the phenomenon of technological mediation has remained an underdeveloped theme, while the postphenomenological framework to investigate technological mediations is in need of methodology and expansion. How, then, to develop a theory of mediation that makes possible an empirical-philosophical conceptualization of the role of technology in the relations between human beings and reality?

ONE MORE TURN AFTER THE MATERIAL TURN

Until recently, and for good reasons, the postphenomenological approach has invested most of its energy in analyzing the relations between humans and technologies, and the mediating roles technologies play in the relations between human beings and the world. One of the central ideas in the development of the postphenomenological approach, after all, is that we need a "material turn" in philosophy of technology (Verbeek 2005; Ihde 1993): we also need to study *things*, rather than merely focusing on *humans*. By now, it is starting to become visible that this focus on materiality also has its limitations. In the "material turn," the ways in which human beings *appropriate* technological mediations tends to remain underexposed. Mediation theory, therefore, needs one more turn after therefore material turn. Having shifted its focus from human understandings of technology toward the materiality of technologies, we now have to move toward to technologically mediated human beings. In order to understand how technologies mediate knowledge, morality and metaphysics, we should not only study technologies, but also the ways in which human beings give meaning to their mediating roles.

The empirical method of Conversation Analysis (CA) offers a highly interesting possibility to investigate this process of "interpretive appropriation," by studying the verbal and nonverbal interactions between human beings in relation to technologies. Conversation Analysis studies how, in the interactions between human beings and their environment, a meaningful "world" is constructed (Edwards & Potter 1992; Edwards 1997; Hutchby 2001; Lamerichs & Te Molder 2011). Just like Latour's Actor-Network Theory (ANT) and Ihde's postphenomenology, CA has its basis in phenomenology and ethnomethodology. But where ANT and postphenomenology have tended to downplay the specific role of the subject in human-technology relations, or even refuse to think in terms of subjects and objects, CA makes it possible to bring the mediated subject to the center again—albeit in an a-modernist and nonhumanist way, because of its integration in a postphenomenological framework. By studying how human beings make technologies epistemologically, ethically, and metaphysically relevant in conversations and interactions, it becomes possible to understand more of the dynamics of human-technology relations, and the interplay between technological mediations and human appropriations.

The method of CA, therefore, adds a dimension to the postphenomenological approach that was missing in order to arrive at a theory of technological mediation. Moreover, CA can also be seen as the missing link between the empirical approach of Science and Technology Studies and the philosophical approach of philosophy of technology: it provides a solid empirical basis for further philosophical analysis of the ways in which technologies help to shape the various dimensions of the relations between humans and reality.

In order to develop a theory of mediation, moreover, the postphenomenological approach should connect itself explicitly to the main subfields of philosophy. A good start would be to focus on *three dimensions* of the relations between human beings and reality: the dimensions of knowledge, ethics, and metaphysics. These dimensions loosely correspond to the three questions that Immanuel Kant saw as the most central ones in philosophy: (1) What can I know?; (2) What ought I to do?; and (3) What may I hope for? The third question will be taken quite freely and broadly as a question about transcendence and metaphysics: how to understand human experiences of what is beyond the grasp of human making and understanding, and what are the most fundamental concepts with which to understand the world?

Connecting these three questions to the phenomenon of technological mediation does not only make it possible to expand our understanding of the social and cultural roles of technology, but also to establish closer connections between the philosophy of technology and other subfields of philosophy. Studying mediation along these lines will shed more light on the constitutive role of technologies in the epistemological, ethical, and metaphysical dimensions of human experiences and practices.

But rather than following Kant's transcendental method, a postphenomenological theory of mediation has an "empirical-philosophical" character, connecting to and expanding the "empirical turn" that has played an important role in the philosophy of technology over the past 15 years (cf. Achterhuis 2001). Empirical-philosophical work does not base itself on the philosophical tradition and conceptual analysis only, but takes the study of actual practices and situations as a starting point for philosophical analysis (cf. Mol 2002). Rather than applying pre-given philosophical theories to technology, it lets technologies challenge existing frameworks, using empirical work as a basis for developing new concepts and frameworks.

In order to develop a theory of technological mediation, therefore, we need to investigate how and to what extent actual technological mediations of knowledge, morality, and experiences of transcendence urge us to develop new conceptual frameworks. And for this, we need to expand the work that has been done in postphenomenological research over the past years and augment it with an empirical dimension.

EPISTEMOLOGY

A first line of research focuses on the epistemological dimensions of technological mediation: how do technologies help to shape our ways of knowing and understanding the world? In Science and Technology studies, processes of knowledge production have always been a central theme, and various empirical approaches have been developed to understand the dynamics of these processes, of which ANT has become the most influential. The role of technological mediations in these processes has remained underdeveloped, though.

To be sure the concept of mediation does play an explicit role in ANT, to indicate that entities are no neutral "intermediaries" but active "mediators" that change situations, relations, and processes. But, unlike postphenomenology, ANT does not explicitly address the hermeneutic dimension of mediations: the ways in which technologies actively help to shape human interpretations of the world. ANT studies processes of mediation "from the outside": as interactions between agents that can be empirically observed and analyzed. Its explicitly symmetrical approach to human and nonhuman agents makes it impossible for ANT to address processes of mediation "from the inside": from the perspective of the experiences and interpretations of human beings who have a technologically mediated relation to the phenomena they are studying. The postphenomenological approach does not accept this radical symmetry—even though it does not accept the modernist subject/object split either.

In postphenomenological theory, the core epistemological idea is that technologies help to shape the reality of the phenomena that are being studied. Technological mediation is no phenomenon that takes place "between" a pre-given world of objects and pre-given human subjects. Rather, human beings and their world are constituted in the "act" of mediation. The epistemological consequences of this relational metaphysics have been labeled as "instrumental realism" (Ihde 1991) and "material hermeneutics" (Ihde 1998). Scientific instruments mediate the reality of the world of scientists; our hermeneutic understanding of the world takes shape in interaction with the material objects that mediate our relation to the world. The content of scientific knowledge and human understandings of the world cannot be separated from the mediating technologies that help to shape this knowledge and understanding.

Other authors have thematized the role of instruments in the development of scientific knowledge as well. The work of Peter Galison (1997) and Davis Baird (2004) also investigates how instrumentation contributes to scientific knowledge—ranging from measuring devices to imaging technologies. Still, they do not explicitly analyze the dynamics of the mutual constitution of "subject" and "object" in the mediated relations between scientists and phenomena. For this reason, Don Ihde's "material hermeneutics" forms an important addition to the—equally important—innovation that ANT brought to epistemology (Latour 1987), and to existing analyses of the roles of instruments in scientific practices. While ANT opened the possibility to study the constitution of realities in empirical ways—remotely based on the phenomenological-anthropological approach of ethnomethodology—postphenomenology added a hermeneutical perspective to empirical philosophy. Rather than studying the networks of relations between actants from a quasi-external standpoint, postphenomenology makes it possible to address processes of mediation "from within": in terms of perceptions and interpretations that can help to shape intentions and actions.

Yet, in order to develop a theory of epistemological or hermeneutic mediation, we need to make one step further. After having opened an empirical way to study the mediated character of the human understanding of the world, and after having articulated the hermeneutic structure of this mediated relation, it is now time to study empirically the content of this hermeneutic structure. How do technologies help to shape the reality that scientists are studying, and that human beings are experiencing in their daily lives? And how do human beings make sense of this technologically mediated reality?

Conversation Analysis provides the missing link here. CA makes it possible to study empirically how the mediating roles of technologies help to shape specific understandings of reality, and how specific "subjects" and "objects" of knowledge are constituted in their technologically mediated relation. While postphenomenological research until now has focused on understanding the mediating roles of technologies, CA makes it possible to investigate how human beings who use scientific instruments give meaning to the mediating roles of technologies and to the mediated realities they help to constitute. By studying the conversations and practices around mediating technologies, we can gain more insight in the ways in which users take up with technological mediations; how they develop creative ways of appropriating these mediations and integrate them in existing practices and interpretive frameworks; how they see themselves in relation to the mediating technologies and to the phenomena they are perceiving or studying. CA then functions as an "empirical hermeneutics": an instrument to study processes of mediation "from within."

Interesting case studies here are brain imaging technologies and antenatal diagnostic technologies. How do neuroscientists arrive at knowledge about the brain on the basis of brain imaging technologies, and how is this knowledge of the brain subsequently translated to knowledge about human functioning—such as the effectiveness of educational programs or systems, the impact of psychotherapy and psychopharmaca, or the development of adolescents? The method of CA can help to study the mediating roles of technology in the constitution of the brain, the creative appropriation of these mediations by neuroscientists, and the ways they translate brain images into claims about human functioning. In the realm of antenatal diagnostic technologies (fetal imaging technologies, amniocentesis, the recently developed blood test for Down's syndrome, pre-implantation genetic diagnostics), interesting questions concern the ways in which these technologies help to constitute the reality of the fetus and the relations between fetus, expecting parents, and health care professionals.

ETHICS

A second line to develop further is the mediated character of morality. In my own work, I have put much effort in thinking through the ethical dimensions of the phenomenon of technological mediation. In my recent book *Moralizing Technology* I have developed the concept of "moral mediation" to analyze how technologies mediate moral actions by helping to shape specific practices and experiences. The paradigmatic example in this study was obstetric ultrasound and its role in moral decisions regarding abortion. Ultrasound constitutes the reality of the fetus in a radically new way, which brings new ethical questions and responsibilities, and reorganizes ethical practices and decisions. From such a postphenomenological approach, moral agency cannot be located in human beings only, but also includes technological artifacts.

This exploration of the phenomenon of moral mediation was only a first step, though. While the focus in *Moralizing Technology* was primarily on the question to what extent technologies can be described in terms of moral agency, we now need to proceed to build a theory of technologically mediated morality. In all attention for the empirical study of *technology*, the empirical dimensions of the ways in which *human beings* take up with technology has remained underdeveloped. The recent revival of "good life ethics" in the philosophy of technology (Borgmann 1984; Brey 1998; Swierstra 2000) and the increasing attention for "subjectivation and technology" (Rose 2006, Brenninkmeijer 2010, Dorrestijn 2012) requires a closer empirical investigation of the normative relations humans have with technologies. How can the complex interplay between the technological disclosure of reality, the technological mediation of practices, and human moral actions and decisions be understood? And what does this imply for the ethical assessment of technologies, and for the ethics of technology design?

The *objectives* of such an investigation, therefore, are threefold. First, it is important to *understand* the dynamics of technologically mediated morality, by developing a framework to conceptualize how technologies mediate moral actions and decisions, and how human beings actively develop a relation to these mediations. Second, the aim is to develop a practical-ethical framework for *assessing* technologies, by identifying and analyzing the ethical questions and issues regarding technologies. And third, it is important to develop a tool for *designing* effective and morally justifiable mediations into technologies.

Understanding and Evaluating Moral Mediation

Within the postphenomenological approach, some work has been done already on the phenomenon of moral mediation. This work either focuses on the technological mediation of morality (Verbeek 2011) or on the processes of subjectivation in which human beings "design their lives" in interaction with technological mediations (Dorrestijn 2012). In both approaches, however, two crucial aspects of the ethical dimension of technological mediation remain underdeveloped. First, the dynamic interactions between technological mediation and human appropriation requires more (empirical) investigation. How do technologies help to shape moral actions and perceptions, and how do human beings take up with these moral mediations? Second, the consequences of the phenomenon of moral mediation for the ethical assessment of technologies needs to be thought through. CA, again, can help here to investigate how technology users perceive and critically appropriate these moral mediations, and how the "everyday morality" regarding new technologies takes shape.

Two controversial technologies can help to illustrate this: a chip for gender selection and the Google Glass project. The gender selection chip is developed by Loes Segerink, in the lab-on-a-chip group of the University of Twente, headed by Albert van den Berg. It is based on the fact that X-chromosomes are a bit longer and therefore heavier than Y-chromosomes; on the nanoscale this difference in weight is significant and useful for separating the two types of sperm cells. The technology has the potential to become available at a large scale and a low price; therefore, the social effect is likely to be quite substantial if the technology would indeed be introduced. (At present, to be sure, gender selection is forbidden in many countries.) Google Glass is an augmented reality technology that provides users with a "second layer" of information when they wear it: they can get information on objects and buildings they see, and potentially also about human beings, when the technology would be equipped with face recognition technology; they can exchange messages, take pictures and record videos, and surf the internet "in the background" of one's activities.

Both technologies will play important mediating roles in our normative frameworks, as they are likely to have an impact on our responsibilities and values regarding our offspring, and our ways of engaging in the public sphere. At the same time, they are socially controversial, which makes it all the more relevant to investigate the "everyday morality" regarding these technologies to facilitate normative discussions about them, and to inform practices of technology design, use, and policy-making. Interestingly, Google Glass is already there and will be introduced on a large scale in 2015, while the gender selection chip is still in development, which makes it possible to investigate how normative reflection could be integrated in the development and, possibly, the social introduction of the technology.

In order to investigate this dynamics of moral mediation and appropriation, it is necessary to study how the gender selection chip and Google Glass play a mediating role in people's moral frameworks. By making the gender of our offspring a matter of choice rather than fate, images of masculinity and femininity, parent-child relations, and possible also social structures, will change. And Google Glass will change our perceptions of and engagement in public space, the value we attach to specific forms of communication, knowledge, social relations, etc.

In close relation to this, it is important to investigate how this expanded analysis of moral mediation has implications for the assessment of technologies. After all, when our moral frameworks are not independent from the technologies themselves, this has consequences for the ways we can do ethics of technology. The technologically mediated character of human morality makes it impossible to give ethics the role of an external judge, assessing from outside which technologies are acceptable and which are not. The frameworks of assessment take shape in interaction with the technologies they are supposed to assess. There is no "outside" of technology from which we can ethically reflect on technology; ethics can only take place "from within" our technologically mediated situation (Verbeek 2013). Rather than *assessing* technologies from an external position, therefore, an ethics of technology based on mediation theory should be directed at *accompanying* the intertwinement of humans and technologies "from within," focusing on the quality of human existence-with-technology (Hottois 1996, Verbeek 2011a).

Also in this context, CA makes possible an interesting expansion of the postphenomenological framework. With the method of CA, it becomes possible to study how human beings make technologies morally relevant in their conversations. Rather than starting from pre-giving normative frameworks that can be applied to technology, it starts from identifying and analyzing normative issues that arise in people's "everyday morality." Such analyses can be the basis for a practical method to support ethical reflection on (new) technologies. Often, the "everyday" moral considerations, frameworks, and arguments regarding technologies remain implicit and unspoken. By making them explicit and analyzing them in terms of ethical theories and the theory of technological mediation, people's implicit normative concerns become available for moral discussion and reflection. Rather than assessing technologies in terms of pre-given ethical frameworks, this approach of "ethical accompaniment" generates questions that are articulated "from within," connected to everyday practices of design, implementation, and use. By making people's practical moral concerns explicit, and analyzing them in terms of ethical theory, they become "available" for reflection and discussion to users, designers and policy-makers.

In order to assess Google Glass and the gender selection chip, the central issue is how human beings make these technologies morally significant in their conversations about and interactions with them. How to understand these everyday moral concerns in terms of ethical theory? And how to facilitate ethical reflection of users, designers, and policy makers regarding these technologies?

Designing Mediations

Moreover, beside *understanding* moral mediation in more detail and developing a new method to *assess* such mediations, mediation theory also makes it possible to *design* moral mediations. Technologies are increasingly used to influence people's behavior in desirable directions. This phenomenon is often called "nudging" (Thaler and Sunstein 2008) or "persuasive technology" (Fogg 2003): technologies can "nudge" or "persuade" users to behave in specific ways. In my earlier work (Verbeek 2011) I discussed why this "mild" form of influencing people's behavior can be defended as morally justifiable. But now it is time to investigate how mediation theory, augmented with CA, can be used to *design* effective and morally defendable mediations. How can mediation theory inform design practices in such ways that they can effectively and desirably contribute to the realization of specific moral values?

By answering this question, mediation theory can make a significant contribution to the growing field of value-sensitive design (Friedman 2002). At the same time, though, mediation theory provides a radically different normative account of the relations between users and behavior-influencing technologies than most other accounts. Rather than defending the autonomy of technology users—like the "libertarian paternalism" of the nudge approach and the "transparency" of the persuasive technology approach do—it takes the mediated character of human existence as a starting point, and focuses on enhancing the quality of the (inevitably) mediated character of human actions and decisions.

METAPHYSICS

The third dimension of mediation to be studied in order to develop a theory of mediation is metaphysics, taken in the broadest sense as studying the most fundamental notions and concepts with which we understand the world. This metaphysical dimension can be addressed from two different angles. On the one hand, we need to investigate the mediated character of metaphysics: how do metaphysical frameworks, in terms of which human beings approach reality, take shape in interaction with technology? On the other hand, we also need to develop a metaphysical framework for understanding the phenomenon of technological mediation itself—a "metaphysics of mediation"—and compare it to other metaphysical frameworks, for example, in classical phenomenology and ANT.

Mediated Metaphysics

In order to understand how technologies help to shape metaphysical frameworks, a good starting point can be to focus on the relations between technology and religion, more specifically on the ways in which technologies help to shape human experiences of transcendence—of what is beyond the grasp of human understanding and manipulation. While there has been a lot of discussion about *science* and religion, the discussion about *religion* and technology is much smaller (cf. Borgmann 2003; Szerszynski 2004; Nye 1994). And here, mediation theory can make an innovative contribution.

The analysis that German philosopher Karl Jaspers made of the phenomenon of transcendence can be a starting point here (Jaspers 1932). Transcendence, for Jaspers, is a "mode of being," rather than something that "exists" itself. Besides being as an "entity" ("things" or "objects") or as "existence" (humans or "subjects"), there is also the phenomenon of being itself: there mere fact that entities are there, and that their existence cannot be entirely reduced to our interventions and understanding. All entities and human beings have this dimension of transcendence: part of their being escapes interpretive frameworks and technological interventions. Human existence, for instance, has a transcendent dimension, because it happens to us, rather than being the product of our own intervention; we find ourselves in our own existence, just like we find ourselves in the world in which we live.

In order to study the relations between technology and religion, therefore, we need to investigate empirically and phenomenologically how technologies help to shape experiences of transcendence. Technological manipulation might seem to be at odds with receptivity for transcendence, but in fact the two have a close relationship. Technology plays a profound role in the ways in which human beings experience transcendence—even to the point that technology can be said to provide an explicit answer to transcendence. Dutch philosopher Jos de Mul, for instance, convincingly analyzed how technology, just like Christianity and Greek mythology, can be seen as a way of dealing with fate, with things that are beyond our control (De Mul 2008).

Mediation theory can offer an additional way to study the relations between technology and religion, by investigating how experiences of transcendence always have a technologically mediated character, and how technological interventions can be seen as mediator in our relation to what is beyond the grasp of human understanding or manipulation. Congenital diseases, for instance, move from the realm of fate to the realm of individual responsibility when it becomes possible to intervene technologically in them.

The key question here, therefore, is: how do technologies mediate practices and experiences involving transcendence? To what extent does technology exclude transcendence, and to what extent does it leave room for it? How does technology help to shape experiences of transcendence, and the transcendent character of the human being and of life itself?

A good example here is brain technologies. What role do fMRI imaging and brain activity measurement devices play in religious experience? What does it imply for religious experiences if they can be connected to specific forms of brain activity? How do "brain-reading technologies" like neurofeedback result in new practices of "meditation"? Also in the field of antenatal technologies, religious dimensions abound. How do gender selection and antenatal diagnostic technologies help to shape the ways in which human beings experience life as a "gift" (Sandel 2004) versus as something "makeable"? IVF and antenatal genetic interventions seem to bring the transcendent origin of the human being entirely within human reach. What does this imply for the transcendence of life itself?

Again, combining CA with postphenomenological analyses of mediation can offer new insights here. From a postphenomenological point of view, the central focus could be on the ways in which technologies help to shape experiences of transcendence, while CA makes it possible to investigate how human beings take up with these mediating roles. How does knowledge about "religious brain activity" help to shape religious experience? How do human beings give meaning to the impact of neurofeedback on their daily lives? How do gender selection technologies and antenatal diagnostics reorganize the boundaries between fate and responsibility, between transcendence and makeability?

The Metaphysics of Mediation

Besides studying metaphysics, it is also important to investigate the metaphysical implications of the mediation approach. Mediation theory can offer both an empirical and a metaphysical alternative to existing approaches in the philosophy of technology and Science and Technology Studies. Just like (classical) phenomenology and ANT it aims to overcome the modernist separation of subject and object (Latour 1993). But, just like ANT, it does not see "postmodernism" as a solution, since the postmodern reduction of objectivity to subjectivity in fact merely reinforces the subject-object dichotomy.

Yet, as said, there is a crucial difference between ANT and postphenomenology. While ANT overcomes the subject-object split by giving up on the subject-object distinction and speaking about human and nonhuman "actants" in a "symmetrical" way, postphenomenology holds on to an asymmetrical approach to humans and nonhumans. Yet, this does not imply that it would start from an *a priori* separation of subject and object. Rather, it sees subjectivities and objectivities as the *outcomes* of processes of mediation. Without *separating* subjects and objects, it keeps up a *distinction* between them—and it is precisely this distinction that makes it possible to address technological mediations and human appropriations along distinct yet closely connected lines, and to approach objectivities and subjectivities as the results of mediations rather than their starting points.

Mediation theory, therefore, does not only have epistemological and ethical implications, but also entails a new position in the discussion about the limits of modernism. It is important, therefore, to investigate its metaphysical origins and implications. How to position and understand the postphenomenological approach in relation to the history of phenomenology, including the work of Husserl (intentionality), Heidegger (hermeneutics, nonmodernism), Merleau-Ponty (perception and human-object relations), Harman (object-oriented philosophy) and Latour (amodernism, modes of being)?

REFERENCES

Achterhuis, H. (2001). *American Philosophy of Technology: The Empirical Turn.* Bloomington: Indiana University Press.

Baird, D. (2004). *Thing Knowledge: A Philosophy of Scientific Instruments*, Berkeley: University of California Press.

Barad, K. (2003). "Posthumanist Performativity: How Matter Comes to Matter." In: *Signs* 28:3, pp. 801–831.

Borgmann, A. (1984). *Technology and the Character of Contemporary Life.* Chicago / London: University of Chicago Press.

Borgmann, A. (1995). "The moral significance of the material culture," In: *Technology and the Politics of Knowledge*, A. Feenberg and A. Hannay, eds., Indiana University Press, Bloomington/Minneapolis, pp. 85–93.

Borgmann, A. (2003). *Power Failure: Christianity in the Culture of Technology.* Grand Rapids, MI: Brazos.

Brenninkmeijer, J. (2010). "Taking Care of one's Brain: How Manipulating the Brain Changes People's Selves." *History of the Human Sciences* 23:1, 107–126.

Brey, P. (1998). "New Media and the Quality of Life." *Techné: Journal of the Society for Philosophy and Technology* 3:1, 1–23.

Dorrestijn, S. (2012). *The Design of our own Lives. Technical Mediation and Subjectivation after Foucault.* Enschede: University of Twente (dissertation).

Dusek, V. (2006). *Philosophy of Technology: An Introduction.* London: Blackwell.

Edwards, D. (1997). *Discourse and Cognition.* London: Sage.

Edwards, D. & J. Potter (1992). *Discursive Psychology.* London: Sage.

Feenberg, A. (2001). *Critica; Theory of Technology.* Oxford: Oxford University Press.

Floridi, L. en J.W. Sanders (2004). "On the Morality of Artificial Agents." *Minds and Machines* 14:3, pp. 349–379.

Fogg, B.J. (2003). *Persuasive Technology: Using Computers to Change What We Think and Do*. Elsevier.

Foucault, M. (1990). *The care of the self—The history of sexuality: 3*. London: Penguin Books {1984}.

Friedman, B, P. Kahn and A. Borning (2002). Value Sensitive Design: Theory and Methods. Seattle: University of Washington, CSE Technical Report 02-12-01.

Galison, P. (1997). Image and logic: a material culture of microphysics Chicago: University of Chicago Press.

Hacking, I. (1983). Representing and Intervening, Introductory Topics in the Philosophy of Natural Science. Cambridge, UK: Cambridge University Press.

Heidegger, M. (1951). "Das Ding." In: *Vorträge und Aufsätze*. Pfullingen: Neske.

Heidegger, M. (1969). *Discourse on Thinking* (trans. J. M. Anderson and E. H. Freund). New York: Harper & Row.

Heidegger, M. (1976). "Brief über den Humanismus." In: *Wegmarken. Gesamtausgabe 9*, pp. 313–364. Frankfurt am Main: Klostermann {1947}

Heidegger, M. (1977a). "The Age of the World Picture." In: *The Question Concerning Technology and Other Essays*. New York: Harper & Row. {translation of Die Zeit des Weltbildes. In: Holzwege. Frankfurt am Main: Vittorio Klostermann, 1950}

Hottois, G. (1996). *Entre symboles et technosciences. Un itinéraire philosophique*. Paris: Seyssel.

Hutchby, I. (2001). 'Technologies, Texts and Affordances'. Sociology 35: 2, 441–456.

Ihde, D. (1979). *Technics and Praxis*. Dordrecht, Netherlands: Reidel.

Ihde, D. (1990). *Technology and the Lifeworld*. Bloomington: Indiana University Press.

Ihde, D. (1991). *Instrumental Realism*. Bloomington: Indiana University Press.

Ihde, D. (1993). *Postphenomenology*. Evanston, IL: Northwestern University Press.

Ihde, D. (1998). *Expanding Hermeneutics*. Evanston, IL: Northwestern University Press.

Ihde, D. (2010). *Heidegger's Technologies: Postphenomenological Perspectives*. Fordham University Press.

Jaspers, K. (1932). *Philosophie*. Heidelberg: Springer.

Kaplan, D. (2009). *Readings in the Philosophy of Technology*. Rowman and Littlefield.

Kroes, P. A. and A. W. M. Meijers (2002). *The dual nature of technical artifacts: Presentation of a new research program*. Techné 6 (2), 4–8.

Lamerichs, J. & H. te Molder (2011). "Reflecting on your own talk: the Discursive Action Method at work." In: C. Antaki, (Ed.). *Applied conversational analysis: Intervention and change in institutional talk* (pp. 184–206). Basingstoke: Palgrave Macmillan.

Latour, B. (1987). *Science In Action: How to Follow Scientists and Engineers Through Society*. Harvard University Press, Cambridge Mass., USA.

Latour, B. (1992). "Where are the Missing Masses?—The Sociology of a Few Mundane Artifact." In: W.E. Bijker and J. Law, *Shaping Technology / Building Society*. Cambridge: MIT Press.

Latour, B. (1993). *We have never been modern* (trans. C. Porter). Cambridge, Harvard University Press.

Latour, B. (1994). "On Technical Mediation - Philosophy, Sociology, Genealogy." In: *Common Knowledge* 3, 29–64.

Latour, B. (1999). "On Recalling ANT." In: *Actor-Network Theory and After*, ed. J. Law and J. Hassard. Oxford: Blackwell.

Latour (2005). *Reassembling the Social: an introduction into actor-network theory*. Oxford: Oxford University Press.

Latour, B. (2013). *An inquiry into modes of existence: an anthropology of the moderns*. Harvard: Harvard University Press.

Mol, A. (2002). *The Body Multiple: Ontology in Medical Practice*. Durham, N. Ca., and London: Duke University Press.

Mul, J. de (2008). *De domesticatie van het noodlot. De wedergeboorte van de tragedie uit de geest van de technologie*. Kampen: Klement.

Nye, D. (1994). *American Technological Sublime*. Cambridge, MA: MIT Press.

Pickering, A. (1995). *The Mangle of Practice: Time, Agency and Science*. Chicago: University of Chicago Press.

Potter, J. (1996). *Representing Reality: Discourse, Rhetoric and Social Construction*. London: Sage.

Rose, N. (2006). *The Politics of Life Itself: Biomedicine, Power, and Subjectivity in the Twenty-First Century*. Princeton University Press.

Rosenberger, R. (2008). "Perceiving Other Planets: Bodily Experience, Interpretation, and the Mars Orbiter Camera." *Human Studies* 31 (1).

Rosenberger, R. (2012). "Embodied technology and the dangers of using the phone while driving." *Phenomenology and the Cognitive Sciences* 11:1, pp. 79–94.

Sandel, M. (2004). *The Case against Perfection: Ethics in the Age of Genetic Engineering*. Harvard: Harvard University Press.

Selinger, E. (ed.). *Postphenomenology: A Critical Companion to Ihde*. Albany, NY: State University of New York Press, 2006.

Swierstra, T. (2000). *Kloneren in de polder: Analyse van het maatschappelijk debat over klonen en kloneren in Nederland*. Den Haag: Rathenau Instituut.

Szerszynski, B. (2004). *Nature, Technology, and the Sacred*. Blackwell.

Thaler, R. H., & Sunstein, C. R. (2008). *Nudge: Improving decisions about health, wealth, and happiness*. New Haven: Yale University Press.

Verbeek, P.P. (2005). *What Things Do: Philosophical Reflections on Technology, Agency, and Design*. University Park, PA: Penn State University Press.

Verbeek, P.P. (2008). "Obstetric Ultrasound and the Technological Mediation of Morality - A Postphenomenological Analysis." In: *Human Studies* 31:1, 11–26.

Verbeek, P.P. (2011a). *Moralizing Technology: Understanding and Designing the Morality of Things*. Chicago: University of Chicago Press.

Verbeek, P.P. (2011b). *De Grens van de Mens: Over Techniek, Ethiek en de Menselijke Natuur*. Rotterdam: Lemniscaat.

Verbeek, P.P. (2013). "Resistance is futile: toward a non-modern democratization of technology." *Techné* 17:1 pp. 72–92.

Chapter 14

Don Ihde's Relevance to the Gun Debates

Evan Selinger

The tragic Colorado Batman shooting that took place in 2012 prompted a wave of soul-searching.[1] How do things like this happen? Over at *Wired,* David Dobbs gave a provocative answer in "Batman Movies Don't Kill. But They're Friendly to the Concept." I suspect Dobbs's nuanced analysis about causality and responsibility won't sit well with everyone.

Dobbs questions the role of gun culture in steering "certain unhinged or deeply a-moral people toward the sort of violence that has now become so routine that the entire thing seems scripted." But what about "normal" people? Yes, plenty of people carry guns without incident. Yes, proper gun training can go a long way. And, yes, there are significant cultural differences about how guns are used. But, perhaps overly simplistic assumptions about what technology is and who we are when we use it get in the way of us seeing how, to use Dobbs's theatrical metaphor, guns can give "stage directions."

INSTRUMENTALIST CONCEPTION OF TECHNOLOGY

The commonsense view of technology is one that some philosophers call the instrumentalist conception. According to the instrumentalist conception, while the ends that technology can be applied to can be cognitively and morally significant, technology itself is value-neutral. Technology, in other words, is subservient to our beliefs and desires; it does not significantly constrain much less determine them. This view is famously touted in the National Rifle Association's maxim: "Guns don't kill people. People kill people."

To be sure, this statement is more of a slogan than well-formulated argument. But even as a shorthand expression, it captures the widely believed idea that murder is wrong and the appropriate source to blame for committing murder is the person who pulled a gun's trigger. Indeed, the NRA's proposition is not unusual; it aptly expresses the folk psychology that underlies moral and legal norms.

The main idea, here, is that guns are neither animate nor supernatural beings; they cannot use coercion or possession to make a person shoot. By contrast, murderers should be held responsible for their actions because they can resolve conflict without resorting to violence, even during moments of intense passion. Furthermore, it would be absurd to incarcerate a firearm as punishment. Unlike people, guns cannot reflect on wrongdoing or be rehabilitated.

BEYOND INSTRUMENTALISM: GUN USE

Taking on the instrumentalist conception of technology, Don Ihde, a leading philosopher of technology, claims that "the human-gun relation transforms the situation from any similar situation of a human without a gun" (p. 27). By focusing on what it is like for a flesh-and-blood human to actually be in possession of a gun, Ihde describes "lived experience" in a manner that reveals the NRA position to be but a partial grasp of a more complex situation. By equating firearm responsibility exclusively with human choice, the NRA claim abstracts away relevant considerations about how gun possession can affect one's sense of self and agency. In order to appreciate this point, it helps to consider the fundamental materiality of guns.

In principle, guns, like every technology, can be used in different ways to accomplish different goals. Guns can be tossed around like Frisbees. They can be used to dig through dirt like shovels, or mounted on top of a fireplace mantel, as aesthetic objects. They can even be integrated into cooking practices; gangster pancakes might make a tasty Sunday morning treat. But while all of these options remain physical possibilities, they are not likely to occur, at least not in a widespread manner with regularity. Such options are not practically viable because gun design itself embodies behavior-shaping values; its material composition indicates the preferred ends to which it "should" be used. Put in Ihde's parlance, while a gun's structure is "multistable" with respect to its possible uses across a myriad of contexts, a partially determined trajectory nevertheless constrains which possibilities are easy to pursue and which of the intermediate and difficult options are worth investing time and labor into.

With respect to the trajectory at issue, guns were designed for the sole purpose of accomplishing radical and life-altering action at a distance with minimal physical exertion on the part of the shooter. Since a gun's mechanisms were built for the purpose of releasing deadly projectiles outward, it is difficult to imagine how one could realistically find utility in using a gun to pursue ends that do not require shooting bullets. For the most part, a gun's excellence simply lies in its capacity to quickly fire bullets that can reliably pierce targets. Using the butt of a gun to hammer the nail into a "Wanted" post—a common act in the old cowboy movies—is an exceptional use.

What the NRA position fails to convey, therefore, are the perceptual affordances offered by gun possession and the transformative consequences of yielding to these affordances. To someone with a gun, the world readily takes on a distinct shape. It not only offers people, animals, and things to interact with, but also potential targets. Furthermore, gun possession makes it easy to be bold, even hotheaded. Physically weak, emotionally passive, and psychologically introverted people will all be inclined to experience shifts in demeanor. Like many other technologies, Ihde argues, guns mediate the human relation to the world through a dialectic in which aspects of experience are both "amplified" and "reduced." In this case, there is a reduction in the amount and intensity of environmental features that are perceived as dangerous, and a concomitant amplification in the amount and intensity of environmental features that are perceived as calling for the subject to respond with violence.

French philosopher Bruno Latour goes far as to depict the experience of possessing a gun as one that produces a different subject: "You are different with a gun in your hand; the gun is different with you holding it. You are another subject because you hold the gun; the gun is another object because it has entered into a relationship with you" (p. 179). While the idea that a gun-human combination can produce a new subject may seem extreme, it is actually an experience that people (with appropriate background assumptions) typically attest to, when responding to strong architectural configurations. When walking around such prestigious colleges as Harvard and the University of Chicago, it is easy to feel that one has suddenly become smarter. Likewise, museums and sites of religious worship can induce more than a momentary inclination toward reflection; they can allow one to view artistic and spiritual matters as a contemplative being.

THE BRAVE ONE

The points about guns made by Ihde and Latour are poignantly explored in the 2007 film *The Brave One*. Unfortunately, many critics examined the film through a humanist lens, and bounded by its conceptual limitations, offered damning reviews. Many depicted the movie as a hyperbolic revenge film. All they saw was a gun blazing Jodie Foster playing a character named Erica Bain who copes with a violent assault (that kills her fiancé and leaves her in a three week coma) by moving through one scene after another of gratuitous vigilante violence, using an illicitly acquired 9mm handgun to settle scores and punish criminals that the law cannot touch. A stir was even caused by the following so-called "liberal" remarks that Foster made during an interview: "I don't believe that any gun should be in the hand of a thinking, feeling, breathing human being. Americans are by nature filled with rage-slash-fear. And guns are a huge part of our culture. I know I'm crazy because I'm only supposed to say that in Europe. But violence corrupts absolutely."

The critics failed to grasp a point that Foster herself underscored in numerous interviews. Despite its market-driven name, the film is not primarily about human virtues or vices. It does not try to discern whether there is an essential experience of bravery or cowardice, and the extent to which characters in the film personify such ideals. Rather, it is an existential meditation that centers on what Foster calls a "deeper and scarier" theme. Looking beyond the explicit plot and its correlative bursts of visually disturbing depictions of violence, makes it becomes possible to recognize that the film explores the anti-essentialist thesis that people are not unified subjects, but instead are beings with fluid and renegotiable identities. Especially in the face of trauma, people can abandon old lives and start new ones. In the case at issue, Erica goes from being a woman who lives a relatively disembodied existence—a radio host who collects the sounds of NY city by blending into its background; a minor celebrity who refuses an offer to appear on television by suggesting that she is more of a voice than a seductive face; and a lover who, at the beginning of the film, is visually contrasted with an athletic looking, long-haired, male-nurse fiancé—to a someone who can kill in cold blood without experiencing the quintessential physical sign of remorse, shaky hands.

By depicting Erica's metamorphosis as a shift away from disembodiment that is brought by means other than consciousness-raising or personal affirmation, *The Brave One* challenges the instrumental conception of technology. Erica's transformation is so explicitly and thoroughly dependent upon technological mediation that the audience is led to infer that without the gun, she would be radically debilitated by her beating; her fate would lie in becoming an apartment-bound recluse.

Reflecting on the centrality of technological mediation to the plot, Foster uses phenomenological language and tells the media that the gun "opens up a world" in which Erica is viscerally "materialized" and therein drawn to dangerous situations (e.g., late night trips to a convenience store and subway) where there is an increased likelihood of encountering violence. Since Erica enters these places because of a technologically induced desire, and not because she is deliberately seeking retribution, it may be fitting to consider the gun—as Latour might suggest, through his notion of "symmetry"—one of the "actors" in the film.

To be sure, *The Brave One* is just a movie. It isn't a scientific study and it does feature a character who has come undone. But if philosophers like Ihde and Latour are right, we've got more in common with her than most are willing to admit. And this possibility ups Dobbs's already high metaphorical ante.

NOTE

1. A slightly different version of this chapter originally appeared in *The Atlantic* and is reprinted with permission. Evan Selinger, "The Philosophy of the Technology of the Gun," *The Atlantic* July 23, 2012. Accessed on March 8, 2014. http://www.theatlantic.com/technology/archive/2012/07/the-philosophy-of-the-technology-of-the-gun/260220/

REFERENCES

Dobbs, David. "Batman Movies Don't Kill. But They're Friendly to the Concept." *Wired* July 23, 2012. Accessed March 8, 2014. http://www.wired.com/wiredscience/2012/07/batman-movies-dont-kill-but-theyre-friendly-to-the-concept/

Ihde, Don. *Technology and the Lifeworld*. Bloomington: Indiana University Press, 1990.

Latour, Bruno. *Pandora's Hope: Essays on the Reality of Science Studies*. Cambridge: Harvard University Press, 1999.

Valby, Karen. "Jodie Foster: Unbreakable." *Entertainment Weekly* August 31, 2007. Accessed March 8, 2014. http://www.ew.com/ew/article/0,,20054140,00.html

Chapter 15

Technology and the Environment

Lessons from Fukushima

Junichi Murata

At 2:46 P.M. on March 11, 2011, a massive earthquake, measuring magnitude 9.0, struck eastern Japan. It was followed by a giant tsunami, which swept away people, cars, houses, and even whole communities over a vast stretch of the eastern coast of the Tohoku region.

However, what made the Great East Japan Earthquake truly historic were the accidents that occurred at the Fukushima Daiichi (No. 1) Nuclear Power Plant. Japanese officials assigned to the crisis afterward rated the accident as level 7 on the International Nuclear Event Scale, making it the only nuclear crisis since the 1986 Chernobyl disaster to be assessed so severely.

More than three years have already passed since the accident. However, more than 140,000 people are still compelled to remain in an evacuation area and cannot return to their homes. Because of the high level of radioactivity, nobody can enter the area near the epicenter of the accident, and many issues concerning the causes of the accident remain unclear. In 2012, four reports on the results of investigations into the causes of the accident were made public by four organizations, including the Investigation Committee of the Government and the Investigation Commission of the National Diet of Japan. However, concerning one of the most important points, that is, the question of whether the earthquake contributed to the disastrous accident before the tsunami struck, the reports arrive at different conclusions.

Recently, the focus has been on the problem of a huge amount of contaminated water, which continues to accumulate at Fukushima, because there appears to be a possibility that some of the contaminated water has been leaking into the ocean outside the power plant site, and deep concerns have been expressed not only in Japan but also all over the world.

Given this situation, there remain many things that we can and must discuss. However, in the following I would like to concentrate on one aspect of the accident at the nuclear power plant, and clarify what we can learn from the case of Fukushima from the viewpoint of the philosophy of technology.

What I would like to emphasize is that technology is not considered a closed system constituted of only technological factors in the narrow sense of the term, but must be considered an open system related to and composed of various factors, including social, cultural, and natural environmental factors. In this sense, technology is to be considered inherently multidimensional.

In this chapter, I would first like to show that the case of Fukushima teaches us in a negative way how important it is to take seriously the multidimensional character of technology and, secondly, that many causes of the accident, which have since been pointed out, are based on a lack of understanding of these characteristics of technology. Thirdly, in the final part, I refer to the philosophy of technology of Kitaro Nishida, the most famous philosopher in modern Japan, because I think Nishida's philosophy of technology can be considered a forerunner of the view of technology that I attempt to present in this chapter. Referring to Nishida's philosophy, I hope to show that the discussions in this chapter are based not only on the present situation of philosophy but also on the history of philosophy in Japan.

TECHNOLOGY AND THE ENVIRONMENT

Technology is sometimes understood with the image of a machine, which can function properly everywhere independently of the environment where it is used. Given this image, people tend to understand technology as something that stands in contrast to social, cultural, and natural environments, and has some power that influences and sometimes determines the situation of a social, cultural, or natural environment. *Technological determinism*, as understood in the broad sense of the term, is one of the most popular ways of understanding technology in philosophy, as well as in everyday life.

The Baconian idea of technology as the power of domination and the use of nature for the benefit of human beings is a popular one, but many concepts proposed by great philosophers, such as M. Heidegger's "Ge-stell" and M. Horkheimer's "domination of instrumental reason," can be interpreted in this way. Even in the field of environmental ethics, this way of understanding is dominant, as demonstrated in discussions on anthropocentrism and anti-anthropocentrism. Technology, understood as the power of human beings, is contrasted sharply here with the natural environment.

This way of understanding the concept of technology began to change in the 1970s, when a new approach in technology studies called the *social constructivist approach* appeared in the sociology and history of technology, and concentrated on a concrete micro-level analysis of the developmental process of technology (Bijker et al. 1987). The influence of this new approach was not restricted to the fields of sociology and history, but extended to discussions on the philosophy of technology.

On the basis of this social constructivist approach, philosophers are now able to look inside technology, which had long been considered a closed black box, and find there not only technological factors but also various social factors, which relate to economics, politics, culture, and values. As the social constructivist's analysis of the developmental process of bicycles impressively shows, the developmental process from technological design and production to use and diffusion of technological artifacts is not determined by a single factor of technological rationality or efficiency, but is open to various factors originating in various fields.

For example, in the first phase of the developmental process of bicycles at the end of the nineteenth century, the design underwent a great change. The first popular model was called the Penny farthing, whose front wheel is much larger than the rear wheel, and which was preferred mostly by young men who enjoyed traveling at high speeds. After a variety of designs had been attempted, a model called Lawson's Bicyclette became dominant. It was the forerunner of the present popular model, which everyone, including women, can use in everyday life. In this way, the meaning of a bicycle is constituted of multiple factors, including the interests of social groups, and in this sense technological artifacts must be considered to have interpretative flexibility.

On the other hand, we must be careful in that social constructivists do not claim society determines technology in the reverse way to technological determinists, who claim that technology determines society. Rather, it is emphasized that a society without technology is impossible just as technology without a society is unthinkable. Technology and society are two sides of the same coin and constitute a complex system that can be called a socio-technical system or a socio-technical network.

This characterization of technology can be found also in the so-called "*Actor-Network Theory*," which illustrates this characteristic more clearly. According to this view, the socio-technical network is composed equally of both human beings and various other factors including artifacts and natural beings.

On the basis of this way of seeing the relationship between technology and various other factors, I would like to emphasize the *interactive* relationship between technology and its environment. While technology influences and constitutes the social, cultural, and natural environments, various environmental factors influence and constitute how technology is developed and realized. Technology and environment are interdependent and closely connected in multiple dimensions.

If we take this interactive relationship between technology and its environment seriously, we cannot simply say that a technology, which is transferred from one environment to another, remains the same. Indeed, as Lynn White, a historian of technology, explained, while in the late Medieval period of Europe windmills became important as power plants, "In Tibet windmills are used only thus, in the technology of prayers; in China they are applied solely to pumping or to hauling boats over lock-sides, not for grinding grain; in Afghanistan they are engaged chiefly in milling flour" (White 1962: 86). Neither can we simply say that nuclear power plants constructed on firm and stable ground and those constructed in an environment where earthquakes occur frequently are the same technology. Without serious efforts at translation, no technology can be successfully transferred from one environment to another.

With regard to this multidimensional character of technology, the process of introducing the power plant in Fukushima should be regarded as fundamentally problematical. In the mid-1960s, the first unit of the Fukushima Daiichi Nuclear Power Plant, which was originally developed by General Electric (GE) in the US, was introduced mainly at the initiative of GE. During the introduction process, an earthquake-resistant design was added, considering earthquake standards in Japan at that time. But, as the Investigation Commission of the National Diet indicates, it is questionable whether this additional design was sufficient. In any case, although various measures were taken to improve resistance to earthquakes and tsunamis, the basic framework of the original design was never questioned. That meant people continued to think that the core technology of a nuclear plant could function independently of environmental factors.

In this respect, it is important that the Investigation Committee of the Government pointed out the lack of a *complex disaster* viewpoint as the fundamental cause of the accident. It is also interesting that in addition to the lack of a *complex disaster* viewpoint, the report emphasizes that it is necessary to face the reality of natural threats with humility and to keep in mind that Japan has had many in its long history (Executive Summary of the Final Report 2012: 28).

But, when it comes to the concept of a *complex disaster*, we cannot forget that a Japanese geophysicist, Torahiko Terada, had already emphasized this characteristic of a disaster in his well-known essays in 1930s. In one of his essays he writes:

> There is one important aspect that we must consider here, something that is easily forgotten. It is the fact that the further civilization progresses the more severe is the damage caused by the violence of nature. . . . It is exactly the people of a civilization who attempt to make disasters all the more severe by accumulating potential energy, which is transformed into the kinetic energy in disasters. (Terada 1934/2011: 10f)

Recently, a historian of technology, Sara Pritchard, proposed the interesting expression, *envirotechnical* disaster, to understand accidents involving a complex system, which encapsulates and foregrounds the dynamic imbrications of natural and technological systems. She wrote: "Continued confidence in the ideal of wholly discrete environmental and technological systems is an unfortunate, even dangerous illusion that can only entrench high-modernist hubris and the belief that humanity and technology are separate from the environment" (Prichard 2013: 129).

If we take this complex *envirotechnical* character of a disaster seriously, disasters cannot simply be differentiated as either man-made or natural disasters. Just as the concept of technology is multidimensional, the concept of an accident and a disaster must also be considered to be multidimensional.

TECHNOLOGY AND THE ACCIDENT

If a socio-technical network is multidimensional and results in complex behavior that is determined neither by technological nor various environmental factors alone, this network shows the characteristic of under-determination, because there is no guarantee that it constitutes and maintains a harmonious and stable unity under various circumstances. It is well known that technology always brings about unintended consequences during the process of its development and use. This characteristic becomes conspicuous in the process of technology transfer, but it can generally be seen everywhere technology is developed and applied. Edward Tenner expresses this unpredictable and unmanageable aspect of technology with the interesting expression "technology bites back" (Tenner 1996). While this characteristic can sometimes be regarded as an origin of creativity, it is at the same time the origin of failures and accidents involving technology.

The clearest cases are various large-scale accidents involving high-risk technological systems, such as accidents at large-scale chemical factories (Union Carbide chemical plant in Bhopal in India), accidents involving space shuttles (Challenger and Columbia), and accidents at nuclear power plants (Nuclear Power at Three Mile Island, Chernobyl, and Fukushima). Charles Perrow, a sociologist, claims that complex technological systems, in which many factors are inseparably and closely connected, always raise the possibility that accidents will occur in an unpredictable, inevitable, and incomprehensible way, and calls them "normal accidents" (Perrow 1999). Nowadays, we are confronted with various problems related to accidents of this kind. It is already almost 30 years since Urlich Beck proposed the provocative concept of a "risk society" to illustrate our present situation.

Until now, there have been various attempts, including Perrow's normal accident theory, to explain and understand this kind of accident. However, we must be careful here, because understanding "normal accidents" means none other than understanding something that includes factors that cannot be understood; and, in this sense, it means an attempt to understand accidents involving high-risk technologies implies something paradoxical. I think this is one of the reasons we should separate an investigation of the cause of an accident from the (legal) question of who is responsible for an accident.

If we are obliged to determine who is responsible, we must explain the process of occurrence with a precise causal relation, which must be understandable as if it were predictable and evitable. However, this causal story has meaning only in hindsight. In contrast, people confronting an accident in real-time must judge everything in an uncertain situation without hindsight to help them. In this sense, the post hoc causal story cannot but neglect and eliminate incomprehensible factors; therefore, it is difficult to learn from it and acquire helpful hints for the future.

Understanding a "normal accident" is nothing less than understanding it as an occurrence that includes some incomprehensible factors. Perhaps one might think that this characterization is dubious. However, this paradoxical characteristic of technological accidents has been well known for a long time. At the beginning of his *Dialogue concerning Two New Sciences,* Galileo Galilei showed impressively that events happen contrary to expectations; in particular, a precautionary measure can have a disastrous result (Galileo 1914: 5).

> A large marble column was laid out so that its two ends rested each upon a piece of beam. A little later it occurred to a mechanic that, in order to be doubly sure of not breaking in the middle by its own weight, it would be wise to lay a third support midway. This seemed to all to be an excellent idea; but the sequel showed that it was quite the opposite, for not many months passed before the column was found have cracked and broken exactly above the new middle support. (Galileo 1914: 5)

Of course, after the unexpected event has happened, we can identify a cause that makes our surprise vanish. However, such a post hoc causal story always arrives too late and cannot be helpful in making understandable the characteristics of an uncertain situation in which people are involved, and the important point that a precautionary measure can have a disastrous result.

Recently, the *Report of the Columbia Accident Investigation Board* of NASA indicated that changing explicit rules and institutional structures is not sufficient to avoid future accidents, because the changes always have the potential to produce new risks. According to the Report, a fundamental change of attitude in the cultural dimension is required. According to NASA's Report:

> The [Space Shuttle] Program must also *remain sensitive* to the fact that *despite its best intention*, managers, engineers, safety professional, and other employees can, when confronted with extraordinary demands, act in counterproductive ways. (Report 2003: 181, emphasis mine)

> Organizations that deal with high-risk operations must always have a *healthy fear of failure*—operations must be proved safe, rather than the other way around. (Report 2003: 190, emphasis mine)

This suggests clearly where we should search for the resources to avoid future possible accidents. They are not found in ethics in the narrow sense of the term, because with the "best intentions" people cannot contribute to preventing failures. Rather "sensitivity" to possible accidents and "a healthy fear of failure" must play decisive roles. This recommendation of the investigation board of the Columbia accident seems to show that it accepts the paradoxical character of "normal accidents" and finds a way to respond to this paradox. If unpredictable accidents are inevitable with technological systems, it is not sufficient to consider only explicit factors within systems. Rather, we must take into consideration unknown factors that lie outside the explicit and rational understanding of systems. The case of Fukushima shows how difficult it is to understand these circumstances. Confronted with the disastrous results of the severe earthquake, tsunami, and nuclear power plant accident, many experts used the phrase "beyond imagination" or "beyond prediction." Meaning that what happened could not have been predicted.

In this respect, the disastrous accident at Fukushima Daiichi (No. 1) Nuclear Power Plant is to be considered a typical case of a "normal accident." It can be characterized as "normal" because it was only clarified after the accident that the people who were involved in developing and operating the power plant continued to think that the multiplex protection system of the power plant was sufficient to avoid a possible severe accident, and did not learn from the accidents at Three Mile Island and Chernobyl. Until the accident occurred, they continued to assume that a severe accident would never occur at a nuclear power plant in Japan, and missed opportunities to improve safety measures when recommendations and criticisms were raised in various fields.

To illustrate these circumstances, almost everyone began to use the term "safety myth," which is to believe in the safety of nuclear power plants without taking seriously the possibility of accidents, and this ensnared people related to nuclear power plants for a long time prior to the accident.

The myth played an important role in the history of nuclear power plants in Japan. For example, in the mid-1990s, an accident occurred at the Monjyu fast-breeder reactor in Tsuruga, Fukui prefecture, which had a serious impact on the public at that time, and led them to mistrust nuclear technology. In the face of this situation, the Nuclear Safety Commission issued a White

Paper, in which it emphasized the need to recover people's trust in nuclear technology, and spread and establish *anshin* (feeling of safety) among the public. They did not think that it was more important to improve the safety of nuclear technology itself, because they thought if they attempted to improve safety measures people would tend to think that nuclear technology had not been sufficiently safe, increasing their mistrust in the technology. It was considered important not to improve safety but to promote a feeling of safety among the public in order to establish nuclear power plants in Japan. In fact, the word *anshin* has since come to be used often in public, not only in the nuclear technology field but also in other fields. This story shows one of the typical ways the "safety myth" played a social and cultural role in Japan.

What is important and surprising now is that this perverse structure related to the "safety myth" seems to remain after the accident. Even now, we often still hear people use the word *anshin* (feeling of safety) together with *anzen* (safety) in various fields, as if saying *anshin* were as important as *anzen* itself. I think this is exactly where philosophy must intervene to break the spell of this myth and clarify what we can and should learn from Fukushima.

A JAPANESE PERSPECTIVE: NISHIDA'S PHILOSOPHY OF TECHNOLOGY

In the previous section, I attempted to show that to understand the meaning of accidents associated with technology and to respond to them it is necessary to take cultural factors into consideration. In particular, when it comes to the "safety myth," which is often regarded as the essential cause of the accident at Fukushima Daiichi Nuclear Power Plant, we cannot but refer to the characteristics of Japanese culture. Indeed, after March 11, we heard various opinions, which attempted to clarify the cause and the process of the accident, focusing on Japanese tradition and culture. One conspicuous example is the statement of Kiyoshi Kurokawa, the chairman of the Investigation Commission of the National Diet of Japan. Kurokawa characterized the accident at Fukushima with the expression "a disaster 'Made in Japan'" and maintained that the essential cause of the accident is rooted in Japanese culture.

> What must be admitted—very painfully—is that this was *a disaster 'Made in Japan.'* Its fundamental causes are to be found in the *ingrained conventions of Japanese culture*: our reflexive obedience; our reluctance to question authority; our devotion to 'sticking with the program;' our groupism; and our insularity. (Kurokawa 2012: 9)

In his message, Kurokawa goes on to state that this mindset can be found across Japan, so everybody in Japan should reflect upon their own responsibilities.

Perhaps, this way of characterizing the cause of the accident seems to be, at first sight, very clear and understandable, because it uses stereotypical expressions to characterize a mindset of Japanese people. However, I think this way of characterizing the cause of the accident is too one-sided and too vague to allow meaningful lessons to be drawn from the accident.

First, in the official report, Kurokawa emphasizes that the accident at Fukushima was "a profoundly man-made disaster—that could and should have been foreseen and prevented" in contrast to the earthquake and tsunami, which were natural disasters (Kurokawa 2012: 9). It is understandable that, as chairman of the investigation commission, he attempts to clarify the cause of the accident and identify who or what was responsible for it. However, if the investigation is directed only in the direction of a post hoc causal story and concentrates on pursuing the factors responsible, the most important characteristic of the accident, that is, the multidimensionality of the accident, is easily neglected. Second, Kurokawa mentions the ingrained convention of Japanese culture, namely our reflexive obedience, our reluctance to question authority, and our devotion to "sticking with the program," which he considers to be the essential cause of the accident. However, these factors are found not only in the behaviors of Japanese people. For example, it was exactly these characteristics that the Columbia Accident Investigation Board of NASA identified in the behaviors of the people working in NASA as the essential cause of the Columbia accident. If one wants to point out cultural factors in these cases, one must, first of all, search for factors that belong to the cultures of bureaucratic organizations in general rather than to the culture of Japan or America in particular.

In this sense, it is questionable whether it is meaningful to characterize the accident at Fukushima only by focusing on the Japanese context in this way. On the other hand, we have already seen in the previous section that in Japan there were people, such as Torahiko Terada, who emphasized the complex character of disasters and criticized the clear distinction made between man-made and natural disasters. In addition, when it comes to a complex viewpoint of technology and environment, we can find intense discussions in the investigations of Japanese philosophers, who struggled to clarify the philosophical implications of their own experiences in the context of modern Japan. The most important result of such efforts was Kitaro Nishida's philosophical view, according to which technology and environment are inseparably connected.

In the modern history of philosophy in Japan, the philosophy of technology has never been a dominant stream. In this context, the situation in Japan has been not so different from that in the western history of philosophy. But in the 1930s, when modern Japan began to appear as one of the industrialized countries in the world, many representative philosophers, such as Jun Tosaka and Kiyoshi Miki, expressed their interest in the philosophy of technology and published books, the titles of which include the expression "philosophy of technology." At about the same time, Kitaro Nishida, who was a teacher of Tosaka and Miki, published many articles, in which he attempted to revise and develop his thoughts about the philosophical system, focusing on the interactive structure between human beings and their environments. He attempted to clarify this interactive structure using the concept of technology.

Nishida published neither a book nor an article that directly addresses problems related to the philosophy of technology. However, in several books of *Philosophical Papers*, in which he collected his published papers, he positions the concept of technology at a central place in his philosophical system. According to him, the ultimate reality continued to be the most important issue of his philosophical investigation since his maiden work *An Inquiry into the Good.* While in his first book the ultimate reality was characterized as a world of "pure experiences," in which there is no separation between subject and object, it is now considered the *historical world*, in which subject and object are closely connected, but at the same time are strictly differentiated. The ultimate reality must be regarded as the world, to which we belong, living and working with our bodies in our everyday lives, and in which *poiesis* or creative activities play a central role. The following expresses his main thesis:

> The historical world is a world of production and a world of creation. Production means that we make things. However, while things are made by us, they become entirely independent and move us in a reverse way. In addition, the acts of our making things themselves originate from the world of things. While things and we remain opposite and contradictory, things move us and we move things, and through this process the world creates itself as a contradictory self-identity. *From that which is made to that which makes*, the world moves through active intuitions. We are able to make things because we are productive elements in the productive world and creative elements in the creative world. (Nishida 1949b: 9, emphasis mine)

From this, we can point out several important features of Nishida's view of technology.

First, production does not begin with someone who attempts to design or produce something new. Rather production can only begin from an interaction with something, which is there in the world as a being already made. Technology as a process from the made to the making is an essential feature of an *interactive relation with the world.* In this sense, technology cannot be understood as something that is independent of its environment, but must be considered to be something that is thoroughly embedded within a certain environment.

Second, the process from the made to the making is essentially a creative process. According to Nishida, our real world has features that must be described with contradictory concepts, such as subject and object, one and many, or motion and rest.

Nishida uses an expression "contradictory identity" to describe the creative structure of the ultimate reality understood as the historical world. Various factors, which constitute the world, are closely connected, but they cannot be unified in a harmonious or coherent way; they oppose and contradict each other. Technology, which is considered to be an essential feature of the historical world, is also composed of various factors, which are in the relation of a "contradictory identity."

In this sense, in the case of Nishida, the essence of technology cannot be understood as a unified framework, in which various factors are connected and controlled in a coherent way, as with Heidegger. In the case of Heidegger, as the well-known concept of "Ge-stell" indicates, the essence of (modern) technology is to be understood as a unified framework, in which everything is dominated and controlled in a definite direction. While the world of "Ge-stell" is governed by a single factor of technological rationality or efficiency, and in this sense is one-dimensional, the historical world of Nishida is composed of various contradictory factors, and in this sense is multidimensional.

Because the historical world always has contradictory characteristics, it cannot be stable: it moves incessantly and is always in a transformational process. This transformational process cannot be characterized as mechanical or teleological, because it is not determined causally or planned or produced with a definite purpose, but arises spontaneously through the interactions of subjects and objects, of one and many. The process is creative because a new situation is always incommensurable with the old one from which it was formed.

Third, this creative process has at the same time an aspect of a cognitive process, which Nishida calls "active intuition." This process is characterized as active, because it consists of an interactive process between subject and object or between the making and the made. It is characterized as intuition, because through the interactive process it is possible to consciously grasp reality. According to Nishida, consciousness and cognition are made possible on the basis of the interaction between a subject and an environment, which is characterized as technological. This conscious and cognitive aspect can be found everywhere in our daily experiences, and in the cognitive skills of artisans and artists, or experimental scientists in their specific forms. In this sense, the concept of technology plays a fundamental role in his philosophical system, including ontology and epistemology.

Fourth, in his philosophy of technology, Nishida emphasizes not only the instrumental role of technology, with which our lives are made convenient and stable. Instead, he underscores the role of technology in radically transforming our historical world. According to him, because of this characteristic of technology, our lives are always in a process of self-negating or self-creating; therefore, they are unstable. "Even in the simple process of building a house, things are not given only as material but as something that has a fateful significance for our actions. In every action we stand on the brink of crisis in some way or other. Our world of everyday life is a world of true crisis" (Nishida 1949a: 70).

Building a house is a technological attempt to realize a place where we are safe from various inconveniences and dangers. However, according to Nishida, if we produce artifacts to promote convenience and improve safety, we necessarily bring about new dangers. Safety and lack of safety are inseparable in our historical world. In this sense, the lack of safety is not a feature we can control absolutely and eliminate totally, but rather is a feature with which we must get along, as long as we continue to retain creativity in our lives in the historical world. In the historical world of Nishida, there is no place where "the safety myth" and perfect *anshin* can be achieved.

In this way, although Nishida never explicitly refers to concrete examples of technology or technological accidents, we can find in his philosophy of technology many insightful theses related to the concept of technology, which correspond to our theses of the multidimensionality of technology.

While the statements of Torahiko Terada were repeatedly referred to and cited after March 11 in Japan, nobody referred to Nishida's philosophy of technology. When it comes to the philosophy of Nishida, it is still the norm for almost everybody to refer to his philosophy of religion, implying a relationship with Zen Buddhism. He is still considered a representative philosopher of Japanese and Eastern culture in contrast to Western culture. In this sense, I think it is all the more important for us to bring to mind that it was Nishida who proposed a view of technology that supports our philosophical attempt to respond to the disastrous events on March 11.

At the end of the previous section, I referred to the contemporary situation in Japan, in which the "safety myth" still seems to have adherents, because the word *anshin* (feeling of safety) continues to be used together with *anzen* (safety). I also mentioned the role of philosophy in intervening in the cultural dimension to break the spell cast by this myth. But, how can philosophy tackle such a task in a meaningful way, if philosophy can have only a status that is separate from the Japanese context in which it attempts to intervene. On the other hand, if we can find in the history of Japanese philosophy a philosopher who emphasizes an insight from which the "safety myth" is severely criticized, the philosophical attempt to intervene in the present situation can be interpreted as an attempt based not only on contemporary philosophical discussions but also on the history of philosophy in Japan. If the history of philosophy can be considered to be a reflexive expression of one aspect of Japanese culture, the philosophical criticism of the "safety myth" and *anshin* can be considered not as an intervention into Japanese culture from outside, but as an intervention from within Japanese culture itself.

We must not forget that Japanese culture is not only constitutive of factors that support the "safety myth" and *anshin*, as the chairman of the Investigation Commission of the National Diet of Japan, Kurokawa, indicated, but includes a reflexive expression that severely criticizes such a myth. The history and culture of Japan itself is complex and multidimensional.

PROVISIONAL CONCLUSION

I think all of this indicates that we need to change the way we understand the meaning of technology. Technology cannot be understood as something that can function as a closed system independent of various environmental factors; rather, it always functions as an open system interacting with various factors, which include unknown and unmanageable elements. Just as the question of how to respond to others is of prime importance in all of the ethics of human beings, responding to "the other" is the central task of technology.

REFERENCES

Bijker, W., Pinch, T., and Hughes, T., eds. (1987). *The Social Construction of Technological Systems: New Directions in the Sociology and History of Technology*, MIT Press.

The Executive Summary of the Final Report (2012). *The Executive Summary of the Final Report,* published by Investigation Committee on the Accident at Fukushima Nuclear Power Station of Tokyo Electric Power Company.

Galileo G. (1914). *Dialogue Concerning Two New Sciences*, translated by Crew, H., and de Salvo, A., Dover Publications.

Kurokawa, K. (2012). *The Official Report of the Fukushima Nuclear Accident Independent Investigation Commission, Executive Summary*, published by The National Diet of Japan, The Fukushima Nuclear Accident Independent Investigation Commission.

Nishida, K. (1949a). *Complete Works*, Vol. 8, Iwanami.

Nishida, K. (1949b). *Complete Works*, vol. 9, Iwanami.

Perrow, C. (1999). *Normal Accidents, Living with High-Risk Technologies*, Princeton University Press.

Prichard, S. (2013). "Envirotechnical Disaster at Fukushima," in *Nuclear Disaster at Fukushima Daiichi; Social, Political and Environmental Issues*, ed. by Richard Hindmarsh, Routledge.

Report 2003. *Columbia Accident Investigation Board, Report* Volume 1, August 2003, Government Printing Office.

Tenner, E. (1996). *Why Things Bite Back: Technology and the Revenge of Unintended Consequences*, Knopf.

Terada, T. (1934/2011). "Natural disaster and the defense of a country," in *Natural disaster and the Japanese people* [*TENSAI TO NIHONJIN* (in Japanese)], Kadokawa.

White, L. Jr. (1962). *Medieval Technology and Social Change*, Oxford University Press.

Chapter 16

How Does Technology Alter Sports

Body, Space, and Ethics

Shoji Nagataki

While our bodies, seen as physical objects, belong to the world, they function as what might be called a *lived medium* through which we perceptually experience the environment. Different organisms have structurally different bodies, which means that a variety of worlds are being experienced by them. Even within the humans, those who have been given special trainings, like athletes and artists, may recognize the world in a different way from those who have not. To use a Gibsonian term, they can be seen to pick up distinctive affordances. Instruments and technological apparatuses also help us to extend our bodies and change our experiences. Furthermore, linguistic ability can influence the way we perceive things. These facts suggest that the way we experience the world can be much more complicated than one might think. To illustrate this point, we would like to focus on athletes' bodies and their experiences.

Athletes playing different sport show a great variety of physical movements, which are regulated by the rules of sports, not to mention by physical restrictions. On a professional level, their speed, complexity and elaborateness can be far beyond those of ordinary people. And what is remarkable about their performance is that it is supported, in an essential way, by scientific technology.

The aims of this chapter are:

1. to sketch two contrasting conceptions of the body in modern philosophy,
2. to analyze an ongoing transformation in the conditions of athletes' bodies and sports, and
3. to consider where it might lead us to.

VIEWS OF THE BODY: A BRIEF HISTORY

The Body in the Modern Era, or the Body as a Machine

We start with seeing how the body image has been made and modified since the modern era. As Merleau-Ponty (1960, 227) put it, "[f]or many thinkers at the close of 19th century, the body is a bit of matter, a bundle of mechanisms." This kind of view of the body goes back to Descartes in seventeenth century. He regarded it as a physical thing controlled by natural laws. La Mettrie, a prominent follower of Descartes, says that "[the] human body is a machine which winds by itself, and it is a living image of the perpetual motion" (Mettrie 37). The nature of Cartesian physical things is made explicit in the following description.

> I . . . see whether anything can be known for certain about material objects . . . . I distinctly imagine quantity—that is, the length, breadth and depth of the quantity, or rather of the thing that is quantified. I also enumerate the thing's parts, to which I attribute various sizes, shapes, positions and movements; and to the movements I attribute various durations,. . . (Descartes 2, 469)

The physical has an extension which is quantifiable and amenable to mathematical description and analysis. Descartes believed that this was also true of the human body.

The development and fixation of the image of the body as a kinetic machine was due to the revival of anatomy as well as to the rise of mechanistic philosophy in the early modern age. The anatomical science remained undeveloped in the Middle Era, mainly because surgeons were of socially lower status for a long time (Mandressi 329). However, after the prolonged period of stagnation, "[t]he knowledge of anatomical science became clear and accurate, and its status improved" (ibid. 331). From the eleventh century to the beginning of the fourteenth century, a body of anatomical literature reappeared in Western Europe, including the Latin translation of Galen. The importance of anatomical practice was recognized through those books. It was on this historical background that William Harvey published his momentous book, *On the Motion of the Heart and Blood in Animals* (1628). The knowledge of anatomy rapidly increased in tandem with the idea of decomposability in human body, which promoted a comparison between body parts, such as skeletons, muscles, inward organs, vessels, skin, and those of a kinetic machine. The framework of the mechanical philosophy and the development of anatomy contributed to building the body image as a mobile machine.

According to Descartes, a champion of mechanistic philosophy, "my essence" consists in the fact that "I am a thing which thinks" (Descartes 2, 488).

He divided human into two parts completely, which are myself and my body that has a quantitative property, through his metaphysical meditation. He thought that myself, a subject, can be clearly distinguished from my body, an object of physical nature. My body is characterized as something controllable at will by myself (Merleau-Ponty 1964, 307). Merleau-Ponty referred to the body in a Cartesian sense as "the objective body," or the "impoverished image" of the true version of the body (Merleau-Ponty 1945, 493). It is a mere object, or an accumulation of bones, muscles, and nerves, existent in the physical space (ibid. 123).

Descartes speaks, however, also of the unity of mind and body. This idea was vaguely suggested in his *Meditations*, where he said that the "body which, by a special right, I called my own, belonged to me more properly and strictly than any of the others" (Descartes 2, 485). The thought of anti-dualism would be elaborated in his letter to Elizabeth (Descartes 3, 44–5).

> That's why people who never come at things in a philosophical way and use only their senses have no doubt that the soul moves the body and that the body acts on the soul. They regard soul and body as a single thing, i.e. they conceive their union. . . . Finally, what teaches us how to conceive the soul's union with the body is the ordinary course of life and communication and not meditating or studying things that exercise the imagination.

Summing up, he argued that humans, who are familiar with philosophical thoughts even slightly, live two different types of life: ordinary and metaphysical. In the former, the unity of mind and body seems an indubitable fact to us, while, in the latter, the distinction between them is clear.

Thus, he had two conceptions of the body:

1. the physical or mechanical as an object of possession and control by the mind—the idea of mind-body dualism and
2. the thing which is intricately and inseparably linked to the mind.

Though most philosophers held that the latter does not reflect Descartes' real thought, some, such as Merleau-Ponty and Gibson tried to develop the latter conception of the body.

Phenomenal Body and the Environmental World

"Experience of one's own body runs counter to reflective procedure which detaches subject and object from each other, and which gives us only the thought about the body, or the body as an idea, and not the experience of the body or the body in reality" (Merleau-Ponty 1945, 231). What does "the body in reality mean?" It is the body as we live through. Merleau-Ponty names it "the phenomenal body" in other places (ibid. 123 et al.) and counterposes it to the objective one. The body in reality, or *the phenomenal body*, challenges the Cartesian dualistic way of thinking in two respects.

First, it is an objection against the modern ontology, according to which a dualistic order of minds and things tidily holds, in that the phenomenal body is not exclusively belonging to neither of the mental nor physical realm. He also names the phenomenal body "the natural self" (ibid. 8). The word "natural" refers to the physical property of the body and "self" in this context means the mental self. In sum, this term expresses a hybrid character of the body in reality, which is opposed to Cartesian ontology. The phenomenal body, with its intentionality and semantic relationship with the environmental world, is not decomposable like a machine, and its parts are "dynamically acquainted with each other" (ibid. 269). This point can be illustrated by a musician playing the piano. She does not consciously represent the position of her arms, fingers and keyboards. Her skills do not exist neither in a pure thought nor in the body as a mere physical object, but in the phenomenal body which, as a "mediator of a world," has a semantic relationship with the world. It integrates a spatial arrangement of keyboards into its own intentional and bodily space.

Secondly, it criticizes against the epistemological schema of subject-object. On a Merleau-Pontian conception, the environmental world, along with things within it, is not a mere object for the phenomenal body. It is a *lived* situation, the element which humans as bodily existence and general organisms "dwell in" (Gibson 7). Seen from their perspective, the epistemological schema of subject-object can no longer hold between the world and organisms.

The Athlete's Body or the Body Re-Mechanized

In the previous section, we have seen an alternative view of the body advanced by the French phenomenologist. Now the question is whether the idea of phenomenal body adequately captures the essential features of the body as we experience it or not. Taking skills of a typist and an organist for example, Merleau-Ponty focused on their unconscious, automatic behavior, pointing to the intentionality and subjectivity characteristic of the phenomenal body. However, we should pay due attention to the fact that we usually treat our own bodies consciously when nurturing those skills. Such experiences are, in fact, quite common; when we get hurt, for instance, our phenomenal body turns immediately into the objective one in a Merleau-Pontian sense. In the case of rehabilitation, it is important for patients to be conscious of every movement of their malfunctioning body parts.

The remarks above can be applied to sport training as well. Athletes playing different sports show a great variety of physical movements, some of which can be very unfamiliar to our daily life. When you engage in some sports, you typically have to start as a novice: you have to see your own body objectively; you have to represent your own body consciously in order to produce its ideal movement as suggested by experts; and you have to manipulate your body as an object with its materiality. This aspect of objects begins to press upon us in those cases. In sum, the body as we experienced is "at once phenomenal body and objective body" (Merleau-Ponty 1964, 179). The "objective body and the phenomenal body turn about one another or encroach upon one another" (ibid. 157).

Speaking of sports, it experienced a big change toward the end of the nineteenth century when many thinkers, according to Merleau-Ponty, thought that the body was a bit of matter, a bundle of mechanisms in terms of its materiality. The big change was the birth of competitive sports which was regulated by specific rules and recorded under a certain control. This is the era when "the investment to muscles" got started (Vigarello 170). As competitive sports got popular rapidly, the consciousness of the body grew stronger. The body began to be objectified and toned up by a systematic training process so that its movement could achieve high precision. Since then sports skills and physical fitness have been analyzed from a quantitative viewpoint. The pattern of the *rowing motion of* oarsmen, for instance, was regarded as lawful and systematic (ibid. 177). The effectiveness of training is expressed numerically by measuring the muscle strength, breathing capacity, and the shape of spine.[1] Put simply, the body and bodily skills have been measured, theorized, and "technologized" (ibid.).

The technological innovation extends over sports tools. For instance, the evolution of tennis rackets has made tennis playing much more speedy and powerful than before. Martina Navrátilová, a prominent tennis star from previous years, told that it had become for us to play tennis easier and more technically by technological progress. In football, the official balls used at the World Cup 2006 made a middle range shoot more dynamic. Football shoes used by many of attackers have been trimmed weight, which helps them to play more speedily for a long time.

There are also negative aspects of this tendency. As the body becomes an object of technological manipulation, its physical limits are so extended that athletes may put their health, even their lives, at risk. Many problems seem to be looming before us: particularly those concerning the intervention of biotechnology and engineering in sports, not to mention doping.

As indicated above, despite the claim of phenomenologists that the body is inherently a subject, it has never lost its status of an object in competitive sports. And now, another way of understanding the body is being developed—a picture of the body which can be modified indefinitely by technology. This is, in a way, a reinstatement of the conception de la Mettrie once advanced, the modern theme of "the human machine" (Michaud 441). As Michaud put it, "[w]e cannot delineate the possible range of technological intervention; we cannot be sure which is acceptable and which is not. Then we cannot imagine how does maintain the identity of the self through the body modification" (ibid. 441–2).

We should also point out that there is another mode of intervention of technology in sports: a modification of the space where competitive sports is practiced. In the following sections, I would like to discuss the significance of the changes in those spaces by employing the phenomenological concept of *lifeworld.*

SPORT SPACES AND THE LIFEWORLD

The Scientific Image of the World and the LifeWorld

It is a widely accepted view that natural sciences describe quantitatively the *real* and *objective* world, while the world we experience is subjective and less amenable to quantitative analysis. This view has been predominant in the modern age mainly because of the tremendous success of science and technology. For example, according to natural sciences, since unperceivable physical phenomena like the radiation in the air is measured and recognized by a scientific tool like a Geiger counter, it truly exists in this world. However, phenomenologists, in a sense, turn such a prevalent view on its head: what is real is not the scientifically described world, but the one we directly experience on a daily basis. As Husserl says, the world of direct perception, or human experience, "is always in advance" (Husserl 113) of the emergence of the world natural sciences seek to describe. Merleau-Ponty also writes "the whole universe of science is built upon the world as directly experienced," and "science is the second order expression" of it. "Science has not and never will have, by its nature, the same significance qua form of being as the world which we perceive" (Merleau-Ponty 1945, II–III). Husserl named it "the life-world" (Husserl 158 et al.).

The lifeworld is, by definition, the perceptual intuitive world which does not include all scientific knowledge. The lifeworld ontologically precedes the scientific world, and the latter world is built upon the former one. However, scientific knowledge and results, cultural products, and historical products have already sedimented into the world in which we really live. Our real lifeworld is filled with those products and knowledge. Scientific activities and their idealized products are always streaming into it through sensible-material media (for example, language and products made from scientific results, etc.), affecting it, and finally becoming part of it. We inherit those through the medium of education, etc. This lifeworld, therefore, encompasses scientific activities and its results, that is, the scientifically verified world in its own universal concreteness. We call it the lifeworld in the broader sense. It can be said that scientists premise the lifeworld in the broader sense and make it as the start point of their scientific activities. So, for example, they can understand at a glance specific scientific meanings of measurement equipment and experimental instruments and so on.

As suggested above, "lifeworld" has two—narrower and broader—senses: the narrower one which is opposed to the objective world supposedly explicable by natural sciences, and the broader one which involves cultural and scientific knowledge accumulated through history. The latter functions, in some ways, like a Kuhnian paradigm which provides the basis for scientific research. Though we may belong to the lifeworld in the broader sense, that of the level of predicative consciousness, we also belong to the lifeworld in the narrower sense, that of the level of prepredicative (perceptive) consciousness. The more one is educated, the more one belongs to the lifeworld in the broader sense. Even a scientist may belong to the lifeworld in the narrower sense when she is not immersed in scientific thinking. For what is directly perceived is a fact that the sun rises in the east and sets in the west, that the earth does not go around the sun, and that the ground is immovable. Even after we have learned Euclidean geometry, perceptual space does not become Euclidean for us. In support of this point, we can cite the fact that depth, which Euclidean geometry does not refer to, is inherent in our perceptual space. In sum, we live in the lifeworld both in narrower and broader senses. Which aspect shows up depends on what kinds of interests we have in particular case.

This is the main point of the phenomenological theory of lifeworld. But, the idea of direct perception or perceptual fact is becoming more and more blurred as technology and knowledge of science are continually permeating our everyday life. They are providing us with a new image of the world, and are altering the conditions of the lifeworld. This is particularly obvious in the sport spaces where the most advanced technologies are introduced. In short, reality and scientific images (virtuality) are getting entangled with each other in there.

THE LIFEWORLD ALTERED: APPEARANCE OF NEW SPORTS SPACE

a. The Hawk-Eye

As stated earlier, science and technology has had great influences on athletes' bodies as well as on the spaces in which sports are practiced. Phenomenological conception of the lifeworld and perceptual facts does no longer seem to hold in professional sports spaces, partly because, as we will see later, human judges are being replaced by mechanical ones.

The judge's calls are unavoidably beleaguered with errors and incoherence simply and merely because she is human. For instance, in a football game, it can at times be very hard for a referee to judge whether some particular move is a foul or not. When it comes to diving, her job gets even harder. Football funs become extremely sensitive to her judgment about whether to allow the goal, or whether to call a foul in the penalty box.

Recently a high-tech judgment system has tentatively been introduced for judging the goal-line call. The system, known as the Hawk-Eye, has already been brought into other sports, such as cricket and tennis. For example, the system is designed to track the ball with 10 high-speed cameras set up around the tennis court, and to compile "the data into a display of its most statistically likely path." Thus, "it is an electronically inferred version of the truth rather than the actual truth" (Calder).

The chair umpire and the linesperson in tennis are basically supposed to call *in* or *out*. The chair umpire can overrule a linesperson's call. Moreover, the players are granted the right to bring a challenge to their calls after the introduction of the Hawk-Eye. Players have three times of challenges in one set. If their challenge succeeds—that is, if the player's challenge is supported by the Hawk-Eye—three times still remains. When the game goes to a tie break, one time challenge is added. Big tennis tournaments in the world have introduced this system, with the exception of the French Open, where the games are played on clay courts so that the balls' imprints are clearly remained.[2]

Some players appealed against the chair umpire before introducing this system. John McEnroe was notoriously liable to a fine because of his relentless objection. However his character was accepted to audience and they loved him. His appeal had become a kind of attraction for spectators. Objections like those of McEnroe were dramatically reduced after the Hawk-Eye.

In effect, they leave the final call up to the machine. The moment when a video image is displayed, there is a great round of applause and sigh. The gallery enjoys this tense moment. People share a story that the Hawk-Eye reports facts in the tennis court. Now, the Hawk-Eye has been accepted as the God's eye view. The tennis court becomes a kind of mythological space. Although audience sometimes shared with players' objection before the Hawk-Eyes, players, audience, and even the chair umpire completely obey the machine's call in this space.

Images displayed by the simulation system displays are understood as representing a *true reality* in an unquestioning manner. It should be noted that, unlike a human umpire, the machine can completely reject the appeals from players. "You lost a John McEnroe outburst" (Calder).

b. Between the Real and the Virtual

At the Dubai Open 2007, the match of Rafael Nadal and Mikhail Youzhny, the Hawk-Eye made a mistake. "Rafael Nadal was furious when he believed a ball to be 'out' that Hawk-Eye had called 'in.' He referred to a mark on the court that seemed to prove his point, but the call stood" (O'Callaghan). "Nadal, chair umpire Roland Herfel of Germany and even Youzhny believed that the ball had landed wide after watching the Hawk Eye" (Gomes). "The makers of Hawk-Eye later explained that Nadal could not see the full stretched impact of the ball on the surface, whereas Hawk-Eye could" (O'Callaghan). In other words, they insisted that even a ball which looks like "out" of the court in the perceptual world can be "in" in *the truly real world.*

At the 2007 Wimbledon final between Federer, the defending champion, and Nadal, the Hawk-Eye greatly irritated the Swiss gentleman. The machine seemed to work to Nadal's advantage four times. At the fourth time, which was a game-deciding shot, he challenged the out call. Unhappily for him, the machine called it about 1mm "in." Though he had rarely complained about the umpires, the Hawk-Eye got him extremely angry. "Can we switch it off?" he said to the chair umpire. But he answered "no." He kept complaining to the chair umpire during switching sides. "How was, in the world, that ball 'in'? That system is killing me today."

"Hawk-Eye reports on its own website that the mean error in the position of the tennis ball as measured by its system is 3.6 mm" (Collins and Evans 294). The tolerance, or margin for error does not matter because the human eyes cannot recognize *in* or *out* of a 200 km/h serve in a few millimeters and such an error would be way below the threshold of human perception. This is a view for the Hawk-Eye side. However, the Hawk-Eye itself called 1 mm *in* on the Wimbledon final.

What does *1 mm in* mean on the Wimbledon grass court? How can there be an accuracy of 1 mm about the more or less distorted lines drawn on the damaged court? In the lifeworld, "the edge of a line painted on grass cannot be defined to an accuracy of one millimeter" (Collins and Evans, 301). "Furthermore, tennis balls are furry and it is not clear that their edges can be defined to an accuracy of one millimeter. In short, at Wimbledon there is no such thing as 'in by one millimeter'" (ibid.). The accuracy of *1 mm in* exists not in the lifeworld, but in the virtual world constituted by scientific technology. However what the lifeworld is relatively difficult to define in this context because it is also socially and culturally constructed in some degree. The stronger the influence of technology on the lifeworld becomes, the more complicated it becomes. We should think of *the ambiguity of the lifeworld* in Husserlian sense which we have just mentioned above.

Athlete's Body and Space Technologized

The world is a perceptive and behavioral space for our own body in phenomenological meanings. It regulates our behavior by many kinds of rules, among which some are physical and some are social.

The sport space where sports are practiced is a game space in that it is constructed by conventional rules and of social character. In this space, fair and impartial judgments are demanded in order to decide on the winner. Even if it is practically undecidable for our perceptual ability, making no decision is unacceptable at all. Since decision-making is part of the rules, its absence means that the space is not fully realized. In a tennis court, for example, the ball is supposed to be either *in* or *out* so that the decision is binary. In this way, the game space is of virtual and artificial nature in the first place, which is in striking contrast to the real, fuzzy lifeworld where, in a physical sense, essentially no binary distinctions hold true. This does not, however, mean that the latter is a deficient space. In fact, attempts to impose strict distinctions on the lifeworld lead, in most cases, only to turmoil and tragedy.

It should be noted that technological interventions in athletes' bodies, sport tools, and sport spaces are closely linked and reciprocal; a high-tech enhancement of athletes can sometimes go beyond the referees' ability to judge and, as the Hawk-Eye case exemplifies, sport spaces have to be reconstituted. This tendency seems to continue for some time until we reach some upper limit.

There are yet two other aspects to the technological intervention in sports:

1. The boundary between the natural body of an athlete and the tech-aided, cyborg-like body is becoming blurred.
2. The distinction between the real space and the virtual one is disappearing.

The images produced by the Hawk-Eye are, for the gallery, parts of reality in the life-world. In fact, most of them do not know that the Hawk-Eye images are created with a statistical technique and, therefore, not without errors and glitches. It does not matter for them whether they are real or virtual; what is important is simply whether it is "in" or "out."

CONCLUDING REMARKS: THE FUTURE OF SPORTS

What will the future of professional sports look like? And the future of the athlete's body? Some people might expect that football players who have bodily skills like Play Station's characters would actually appear. Others might envisage a future in which athletes who have technologically enhanced, cyborg-like bodies and robots would play in the Champion's League. Or they might forecast a future in which tennis cyborgs implanted with super small high-speed cameras in their bodies, simulating the ball's trajectory at point-blank range, compete with each other over their advanced skills.

It was a symbolic moment in the history of sports when Oscar Pistorius, the South African athlete with artificial legs, was approved to participate in the London Olympics, 2012. He hoped to run in the Beijing Olympics, 2008, but the International Association of Athletics Federations rejected his entry. He filed an appeal to the Court of Arbitration for Sport, which accepted his suit. The decision of IAAF seems to have a point, because his legs look very different from the usual artificial ones and can reasonably be regarded as an *enhancement* rather than a walking aid. Though his entry was accepted in the London Olympics, the circumstances might have been different if he had been a medal hopeful. At the least, the use of the artificial legs would have been much more controversial. Just like a doping issue, an endless cat-and-mouse game may well continue between regulation and technological innovations, such as in hardware engineering, neuroengineering, and bioengineering. The demarcation between cheating and legitimate enhancement will be likely to present an increasingly serious problem to us.

One thing to point out here is that the decision of legitimacy seems to depend partly on whether or not a story is involved in her performance—the somewhat comforting one that it is a result of day-to-day, back-breaking training. The reason why anabolic steroid, for instance, is regarded as a doping substance is not just that it is harmful to the human body from a medical point of view; the fact that it can help boost muscles in a short time also seems to matter.

So, if there is a long story about painstaking training, do we recognize its legitimacy? Some parents and coaches who want their children or players to be professional athletes sometimes give special supplements and growth hormone to them in their childhood. This process takes much more time than in anabolic steroid cases. But if you hear this kind of body building story, some of you might feel it unnatural or inauthentic. In sum, the demarcation between natural and unnatural is also fuzzy here.

If cyborgs like those appearing in *Ghost in the Shell*[3] were to become common, that demarcation would become fuzzier. This doesn't mean, however, that those technologically enhanced bodies will be accepted straightforwardly in sports or in our daily lives. They can threaten the very meaning of athletic competition, especially if it is readily possible to enhance the athletic ability far beyond that of the usual person. To say the least of it, the acceptance of cyborg-like bodies will require a comforting story that its owner had to undergo a long and harsh training in order to adapt herself to her particular physical condition.

The inexhaustible desire to be faster and stronger sometimes drives people to a hard training in their childhood and doping, including protein and hormonal agent. Judgments in games may be beyond the ability of usual human perception. So, as remarked above, we sometimes leave the decision to technological judging systems, which make the distinction between the real space and the virtual one fuzzy. Though Nadal was much distressed by such a system, he described it as "a positive thing for the sport. The crowd enjoys it and I enjoy because you have your chance [to appeal]" (Calder). We may come to accept the sports transformed by scientific technology, as Nadal suggested. Cyborg-like humans may play the professional tennis in the near future. We may come to say that if the crowd enjoys it, it does not matter.

As time passes, many kinds of enhancement technology may well be accepted and become common in the case of competitive sports. I would not like to discuss here the rights and wrongs of this possible tendency. The best that I can say is that the athlete's body will be much changed in a thinkable future world where cyborgs and designer babies commonly exist.

NOTES

1. The appearance of the dynamometer was the end of eighteenth century and its spreading out was nineteenth century. (Vigarello and Holt 331).

2. This system has not been introduced into the French Open because games are played on clay court where the ball leaves a mark clearly enough for umpires to examine. We sometimes see the scene of a chair umpire stepping down from her chair and checking a mark. However it is also pointed out that "the ball mark can be larger than the compressed ball's footprint because a forceful impact can displace surrounding clay particles" (Fischetti 96).

3. *Ghost in the Shell* is a 1995 anime science fiction film. The world of this film has become interconnected by a vast electronic network that permeates every aspect of life. Much of humanity, including the protagonists, has direct access to this network through cybernetic bodies, or "shells", which possess their consciousness and can give them superhuman abilities (Wikipedia).

REFERENCES

Calder, Peter, "Tennis: Technology takes out doubt and fun", *The New Zealand Herald*, January 30, 2011. http://www.nzherald.co.nz/sport/news/article.cfm?c_id=4&objectid=10702966 (accessed March 1, 2014)

Collins, Harris and Evans, Robert, "You cannot be serious! Public understanding of technology with special reference to 'Hawk-Eye,'" *Public Understanding of Science* 17, 2008, 283–308.

Descartes, René, *Œuvres Philosophiques 3 vols*. Paris: Garnier, 1963–89.

Fischetti, Mark, "In or Out?" *Scientific American* July: 96–7, 2007.

Gibson, James-Jerome, *The Ecological Approach to Visual Perception.* Boston: Houghton Mifflin Company, 1979.

Gomes, Alaric "Hawk Eye leaves Nadal and Federer at wits' end: world no 2 Rafael Nadal has questioned the efficiency of the new Hawk Eye line calling technology", gulfnews.com, May 3, 2009, http://gulfnews.com/sport/tennis/hawk-eye-leaves-nadal-and-federer-at-wits-end-1.165119 (accessed March 1, 2014)

Husserl, Edmund, *Husserliana Bd. VI.* Haag: Martinus Nijhoff, 1954.

La Mettrie, Julien Offray de, *L'Homme machine: Avec une introduction et des notes de Assézat.* Paris: F. Henry, 1865.

Mandressi, Rafael, Dissections et anatomie, in *Histoire du corps 1: De la Renaissance aux Lumières*, ed. Georges Vigarello. 327–50. Paris: Editions du Seuil, 2006.

Merleau-Ponty, Maurice, *Phénoménologie de la perception.* Paris: Gallimard, 1945.

———, *Signes*, Paris: Gallimard. 1960.

———, *Le visible et l'invisible.* Paris: Gallimard, 1964.

Michaud, Yves, Visualisations: le corps et les arts visuels, in *Histoire du corps 3: Les mutations du regard. Le XXe siècle*, ed. Jean-Jacque Courtine. 431–51. Paris: Editions du Seuil, 2006.

O'Callaghan, Jonny, "Top 5 Facts: Controversial Hawk-Eye Moments," January 29, 2011. http://www.howitworksdaily.com/technology/top-five-facts-controversial-hawk-eye-moments/ (accessed March 1, 2014)

Vigarello, Georges, S'entraîner, in *Histoire du corps 3: Les mutations du regard. Le XXe siècle*, ed. Jean-Jacque Courtine. 169–206. Paris: Editions du Seuil, 2006.

Vigarello, Georges and Holt, Richard, Le coros travaillé: gymnastes et sportifs au XIXe siècle, in *Histoire du corps 2: Dela Révolution à la Grande Guerre*, ed. Alan Corbin. 321–88. Paris: Editions du Seuil, 2006.

Chapter 17

Somatology of Aurality

The Voice of the Material World

Eduardo Mendieta

We come to a writer not the way they wanted to be read, nor the way they themselves wrote their corpus. Coming to a writer is like walking through a forest. You may get lost. You may find paths. You may find elevations that give you a sense of distance, of what you have yet to see. But coming to a writer is also very different from how we come to the edge of that forest. A writer writes forward, leaving behind a trace, a layered accumulation. We readers come to that body of layered work from different angles, entering that sedimented history at different times. Take for instance how many of us are introduced to Kant from either the *Critique of Pure Reason* or the *Grounding for the Metaphysics of Morals*, both of which are mature works. We enter Kant's corpus three quarters of the way of its development. Or take how many come to Plato through his *Republic*, which is a middle late work, in which we already have a metaphysically grounded epistemology, ethics, and politics. I have come to Don Ihde's work, backward, reading him from his most recent works to his earliest (Mendieta, 2003 and 2012). It has been fascinating to dig, rummage, slowly work beneath the layers of Ihde's texts, if only because I have been able to see the emergence of a distinct Ihde style, and a unique Ihdean voice. Those of us who have been reading his work from the last two years will be surprised to read his early seventies texts. There is an Ihde that, stylistically, is almost on the opposite spectrum of the clear, transparent, non-metaphysical, highly historical, Ihde that we hear and read today. A comparison immediately suggests itself. Whereas Heidegger's prose and philosophical writing got more obscure, more ethereal, more fragmented, more detached from the world and human experience, as he moved from *Being and Time* to the post-Turn works, Ihde turned away from the quasi-poetry of ontological phenomenology to the unembellished, direct, democratic writer of a philosopher interested in extracting from quotidian language and experience already inchoate understanding and knowledge. This means, to our surprise, that at one time Ihde wrote in a quasi-Heideggerese that I am sure would make him cringe if he heard it again. Here are a couple Ihde Heideggerese pronouncements, from his *Listening and Voice*, first published in 1976 (Ihde, 2007), but written in 72–73:

> The beginning of man is in the midst of *word.* (Ihde, 2007:3)
>
> If philosophy has its very roots intertwined with a secret vision of Being that has resulted in the present state of visualism, can it listen with equal profundity? What is called for is an ontology of the auditory. (Ihde, 2007:15)
>
> Listening begins by being bodily global in its effects. (Ihde, 2007:45)
>
> What remained hidden from my eyes is revealed to my ears . . . Hearing interiors is part of the ordinary signification of sound presence and is ordinarily employed when one wishes to penetrate the invisible. (Ihde, 2007:71)
>
> Emptiness which can be uncanny is silence in the auditory dimension (Ihde, 2007:83)
>
> "Sound dances timefully within experience. Sound embodies the sense of time." (Ihde, 2007: 85) and "Sound reveals time." (Ihde, 2007:102)
>
> "The voice of language domesticates the World." (Ihde, 2007:142) and "Word resides in myself in such a way that language 'lets be' a World as a significant phenomenon." (Ihde, 2007:144)

One last one, perhaps one of my favorite Ihde aphorisms:

> Just as there are voiceless words, there are wordless voices, the voices of things which are a wordless speaking. (Ihde, 2007:154)

Still, even as Ihde appropriated Heidegger's poetic philosophizing, he was also taking tremendous distance from his phenomenological ontology, already back in the early seventies. We know that Ihde has been an acute critic of Heidegger's philosophy of technology. In fact, his *Heidegger's Technologies* has become one of his most popular texts (Ihde, 2010). Yet, I think that Ihde's engagement with the phenomenological tradition in general and with Heidegger in particular goes beyond this type of localized engagement and circumscribed critique. I think that Ihde's success as a technoscience philosopher may have contributed to the neglect of Ihde's equally momentous contributions to phenomenology, in particular, and to what we now call continental philosophy, in general. In the following, therefore, I will focus on Ihde's work as a unique phenomenologist, by profiling what I would call Ihde's critique of Heidegger's bedazzlement. I then want to discuss Ihde's contribution to a truly enfleshed, embodied, fully corporeal phenomenology that does not succumb to the seducements of sight and the delights of the visual, the scopophilia of metaphysics. At the center of this phenomenology of global embodiment is a critique of what Ihde called double reductions of sensibility and perception. I will briefly profile some of these reductions. Then, in a third section I want to briefly bridge the early Ihde of ontologies of sound and phenomenologies of hearing with the later Ihde of technologically mediated embodiments by arguing that an ontology of sound that is in the flow of time must become a somatology of aurality (Mendieta, 2012a and 2014). By this last term I am naming not simple the way we listen, the way we are in an auditory field and the sound of time passing reveals itself to us, but also, and in a co-originary way, how things sound, how things reveal time in their very sounding, in their vibrating and swaying. Things have wordless voices, as Ihde put it, because they sound. The somatology of aurality is attentive to both the sounding of things and their being heard. If to language is to world, then we should say more precisely that sounding is worlding, but sounding is part of a way in which an auditory field constitutes a horizon of temporal presence. All beings have an auditory aura. Being, before it comes to presence in the clearing of truth, is the sound of time. If language is the house of being, Ihde shows, it is because sound is the rhythm of time—and only then can we understand why *Being and Time* and *Time and Being* are interchangeable.

HEIDEGGER'S BEDAZZLEMENT

Ihde' work is unapologetically and unabashedly critical of what he has called in personal conversations "generic continentalism." He has also made it public, at different conferences, that he intends to write a book titled "Against the History of Philosophy." It is not hard to see why the critique of generic continentalism may turn into the critique of the fixation on the history of philosophy that marks so indelibly and unmistakably continental modes of philosophizing. Yet, notwithstanding Ihde's pronouncement, it is also not hard to see to what degree his own work has moved, advanced, and developed through a critical engagement with figures like Husserl, Heidegger, Ricoeur, not to mention Latour, Rorty, Haraway, Harding, Galison, just to mention the most prominent ones. As the expression goes, Ihde protests too loudly, so as to distract us from his own practice. In fact, unwittingly, Ihde's work demonstrates why the continentalist approach to philosophy is superior to the so-called analytical style.

Let me focus on the book that is fast becoming my favorite of his, *Listening and Voice: Phenomenologies of Sound*, in its 2007 expanded second edition. This book, which was written in the early seventies as was already noted, does two things that are distinctly unique to continental ways of philosophizing. It offers a critical reconstruction of the history of philosophy from the perspective of something that has been revealed to have been neglected or occluded by that very tradition. Here, deconstructing the tradition becomes a way to philosophize that illustrates how it is that one can both ask the kinds of questions one can ask, and why new questions can be asked. We can call this, along with Karl-Otto Apel, the *Selbsteinholungsprinzip*, or principle of self-recuperation, which guides continental style philosophizing. In other words, this principle names the very preoccupation of philosophy with its own historicity, and how philosophizing in part has to be about how we have philosophized in the past. Thought has to always give an account of what it used to be, what it has become, and what it ought to become.

The second virtue of the approach I am attributing to Ihde has to do with what I would call "subversive" readings that are not simply formulated extrinsically, as if from above or a *sub specie aeternitatis* standpoint, but rather, they are subversive because they are immanent, from within and given the very assumptions of the tradition. These readings work from within the very assumptions of a particular thinker. Ihde's *Listening and Voice* offers these two types of readings. The first chapters give us a bird eye's view of the way in which sound has been suppressed as visibility has become hegemonic. In fact, Ihde shows how metaphysics itself became possible on the basis of what he called a double reduction: first the reduction of all holistically embodied sensibility to visibility, and second, the reduction of visibility itself to significance. Western metaphysics has been enthralled by the visible that is no longer seen because it is now to be properly captured by the mind's eye. The metaphoricity of the mind's eye, in other words, conceals the way in which seeing is but a dimension of embodied sensing. We are not in the world first as camara obscuras or observing eyes. We come to the world first and foremost through the womb of sound in which we gestate. The world is above all a chamber of sounds, which reverberate through our bodies.

We can say that the occularocentrim of Western metaphysics is grounded in the tyranny of the visual over the democracy of the senses. Evidently, this tyranny is accompanied by the denigration of embodied perception. The type of deconstructive genealogy that Ihde develops in *Listening and Voice* in particular is surely one of the most important ways in which we historicize our present, and thus open ourselves to the fact of the contingency of the constituted world. For Ihde shows how, to paraphrase Heidegger, the history of Western metaphysics is the history of the forgetting of embodied cognition, and above all of our being in the world in and through sound. Western metaphysics is the silencing of the echoing of indwelling.

Ihde's text, however, is doubly striking because of the "silent" and "diplomatic" way in which his reading of both Husserl and Heidegger, and especially of the latter, reveals the ways in which they remained caught within the snares of this double reduction of embodied perception. Ihde's quick but perspicacious reading of Heidegger demonstrates how the language of presence, disclosure, revealing, concealment, and so on, exerted a powerful gravitational pull in such a way that Heidegger himself remains bedazzled by the lighting flash of *Seyn*. Heidegger is blinded by Being that is pure luminosity in the clearing of truth. Presencing, coming to presence, appearing, revealing, all overwhelm the sound of things, and the sounding of being itself. In fact, it is noteworthy that in *Being and Time*, Heidegger claims that: "*Phenomenological truth* (*disclosedness of being*) is *veritas transcendentalis*," (Heidegger, 2010: 36) a statement that subordinates presencing to the revealing of something that is itself not visible. In Ihde's terminology, we can say that at least in the Heidegger of *Sein und Zeit*, there is a double reduction of disclosedness; first to visibility and then to the significance that is not sensible (*veritas transcendentalis*).

What Ihde teaches through his critical reading of Heidegger is that the forgetfulness of being is part and parcel of the forgetfulness of the voice of beings, and in particular of the ways in which sound, as he writes, "reveals time." What we can also cull from Ihde's acute readings is that Heidegger's overt occularocentrism is gained at the expense of the depreciation of sound. This derision or denigration of sound is made manifest in at least two resounding aspects of Heidegger's early work. First, there is the jaundiced look at what Heidegger calls "idle talk," which can be read, notwithstanding Heidegger's own qualifications, as a denigration of everyday speech. The second, has to do with Heidegger's given ontological privilege to ancient Greek and a German that is as remove from everyday as ancient Greek is from modern Greek. In this attitude, Heidegger privileges language not as voice, not as embodied sound, not as the proximity of Dasein to Dasein, not as a co-originary dimension of mit-Sein, but as significance disembodied from language as voice, as the luminosity of appearing.

Ihde undertakes these critiques, not simply to undermine the phenomenological project, not to belittle what phenomenology can accomplish. On the contrary, Ihde's critique of Heidegger's bedazzlement is meant to rescue the tradition from a very evident occularocentrism. Additionally, what makes them even more interesting is that while Ihde is not a "Heidegger Expert," his critique anticipated many of the interesting turns that Heidegger himself would have to make in work that is now regularly been published and translated. Note for instance the second part of the *Beiträge zur Philosophie* (Heidegger, 2012), which is titled "Der Anklang," which has been translated as "Resonating." Now, if we read the first fugues of paragraph 50 of this second section, we encounter the following:

RESONATING

Der Anklang is the echoing of the vibrating of be-ing:
of the essential occurrence of beyng.
out of the abandonment of being
Through the compelling plight
of the forgottenness of beyng (Heidegger, 2012: 85)

Being is rhythm, and the forgetfulness of being is the forgetting of this echoing, vibrating, swaying. I submit that this is what Ihde meant when he wrote: "Sound dances timefully within experience. Sound embodies the sense of time." (Heidegger, 2012: 85). We could then say then that: Language is the house of being, because voice is the house of sounding, and sound is the swaying timing of being.

BODIES AS TECHNOLOGICALLY MEDIATED EMBODIMENTS

Originally, I must confess, I was tempted to call this section Idhe's body. Most of Idhe's texts are populated with descriptions of phenomenological experiments with his body. There is Ihde skiing. There is Ihde cannoning. There is Ihde learning to use reading glasses. There is Ihde learning to walk with his new cyborg knees. There is Ihde and his hearing aids. But, note, there is never a simply naked Ihde body. It is always a body in relation. What is important is that Ihde's body is always a relational body. Ihde's body is a body in interaction with micro and macro-technologies. There is no embodiment that is not always already mediated. To be a body is to be mediated. One could say that each one of us has many bodies (an ideal, projected, assigned, and a lived body), and the task of coming to the world is learning how to bring those disparate bodies into synchrony. A body is always a layered palimpsest of multi-imagined and real, embodied, bodies. But, more than simply negotiating the "ideologies" of bodies, we literary learn to inhabit our bodies through our interactions with many of our culturally produced prosthetics. To come to the world, to be in the world in relation with others, is also and co-originarally, a learning to inhabit a certain type of body. Being human is as Helmut Plessner argued, being an embodied creature that is both a body and has a body (Plessner, 1981). We could supplement Plessner with Ihde and argue that to be human is to negotiate coming to a body that we have, but which we learn to inhabit by negotiating cultural practices of embodiment and by learning to be both uncomfortable or comfortable in it as it ages. These multiple bodies have their own resonances if only because the passing of time echoes in its noises.

I would argue that all of this seems fairly uncontroversial. What I mean is that Ihde has spent most of his philosophical career showing how we are not simple organic bodies, or bodies as naked facts of phenomenological experience. We are "bodies in technology," to use a title of his books, and thus, also bodies of technology (Ihde, 2001). Our bodies are our technologies, and were our technologies to be discovered by a future alien species, after we had become extinct, the general contours, shape, solidity, needs, and vulnerability of our bodies could be extrapolated from those ruins of our technologies. A naïve question can be asked: can we have a direct, unmediated access to our bodies? I think the answer that Ihde has been teaching us to accept is that this is not possible. We are cyborg all the way down, so to say.

Ihde's contribution, however, is more interesting than this. Bodies are practices of embodiment, that is to say, to be a body is not simply to be technologically mediated, but to be mediated by certain embodiment practices. This means, more precisely, that we inhabit our bodies, but in certain ways. So, as we have become *Homo Sapiens*, we have learned and trained our bodies to give privilege to sight, and to attenuate or bracket the power of the other senses. In fact, this learning to inhabit our bodies by either domesticating or suppressing one or many other senses occurs also with respect to one or many senses. So, for instance, through Ihde's phenomenologies of sight and sound, we learn to see that seeing is also domesticating sight, hearing is domesticating hearing in a certain way. We don't see frontally. We learn to see by focusing on what is immediately in front of us, and by focusing on certain visual signals that are culturally marked even before we see. We see what we can see because we have trained to see it. We hear what we can hear because we have been trained to hear what we hear. But we also learn to see, as we see, because we learn not to hear as we see, or to hear only certain things.

The body thus is not simply a technologically mediated practice; it is also a culturally mediated practice that domesticates our senses one way or another. Now, what is interesting in this insight is that here phenomenology meets genealogy. Here Maurice Merleau-Ponty and Michel Foucault converge by way of Ihde by letting us experience how bodies are habits of habitation and not givens, or biological *aprioris*.

SOMATOLOGY AS THE LOGOS OF A HORIZON OF EMBODIMENT

Let me begin this section with a phenomenological story about. my own body. I have been an apprentice runner for nearly three quarters of my life. I began to run when I was in my teens. I used to run through dirty roads, up and down steep hills, through pasture fields, and up the hills of coffee plantations in the mountains of Colombia. Of course, I used to run without a Walkman. I remember when these made their appearance. For a long time I ran without one, mostly because I couldn't afford one, and the idea of running while listening to music seemed weird. Part of the enticement of running was that I would hear the world—the birds, the wind, the trees swaying with their rustling, the sound of my own steps on the gravel, cement or grass. Running was about the sound of the world, as much as about hearing my breathing, feeling the blood engorge my thighs, and hearing the swoosh in my ears of my speed. Eventually, when I moved to New York City to do my graduate work, I did pick up a Walkman, and learned to run with music in my ears. I could now focus on running while having the music set my tempo. In the last few years, since iPods appeared, I almost never run without one. But I have forced myself on occasion to run without one, so that I can get back to setting my own pace, and having a different relation to my body while running. The iPod not only silences the world, but also my own body. I don't hear myself running, and I don't hear the swoosh of my own speed. I bring this example up because I want to foreground the way in which different technological artifacts enable different experiences of sound. The sound difference between a Walkman and an iPod is like that between a vinyl record and a cd, as this last one is distanced from a digital version of a song. The world sounds different when heard through these different embodiment relations.

The world is noisy but not always for the same reasons and in the same ways (Keizer, 2010; Labelle, 2010). Some of us may still remember walking into either university or government offices and hearing the distinct rapid clicking, humming, and the swish of the key ball returning to the beginning of the next line of electric typewriters. I am both old enough and young enough to have lived through that momentous transition, when we were going to migrate to the "paperless office" because of computers. Paper did not disappear, but the sound of offices did change. Offices have become very quiet, although you do hear the softer, muted clicking of keyboards.

The world is noisy because of the way things are made, machines are engineered, and all dwelling spaces are designed in certain ways. Sound Engineering is a major field, and highways are now walled in to divert the sound of speeding cars. Our world has a sound, and it is not simply an indistinct noisiness. A German city has a different sound than an American city, if only because cars made in Germany, or England, sound different than those made in the US. They sound different because they were made to produce a certain kind of affect. It is for this reason that to this different sounding, or "resonating" of the world, we attribute different tonalities. Sound is affect. The world's sounding, the sound of the world, is also accompanied by a certain affect. The world has a mood because it has a tonality. The tonality of the world is its mood. We play certain music or sounds, to get into a mood. But, the world is already in a mood setting sound.

I thus want to bring to conclusion these reflections by nudging Ihde's work on aurality into the direction of what I called here the Somatology of aurality, by which I mean both a phenomenology and a genealogy of the different technologies of embodiment, especially as these technologies affect the way the world sounds, the way we sound, and the way we therefore we hear the world. Thus, for instance, we can begin to talk about the "the gender of sound" (Carson, 1995: 119–142) or the "sound of race" (Mendieta, 2014 and 2012a). If bodies are practices of embodiment, and the body is the trace of a technology, then part and parcel of those practices and technologies is the producing of affect through and in sounding practices. Affect as the tonality of the world also has a prosody, a rhyming, a swaying, a beat. The "prosody of affect," to quote an expression I have used in the past, however, is meditated by our technologies. Thus, our many technological devices, from the telephone, through the television, the iPod, the smart cell phone, skype, to talking cash registers, and automated telephone response voices, we have affected the way we sound and the way we perceive the sounding of others (Appelbaum, 1990; Dolar, 2006; Cavarero, 2005). Coming around full circle to Ihde's critique of the enthrallment of Western metaphysics with sight and the luminosity of being, we can conclude by saying that before we learn to see the world, we learn to hear it. The world reverberates in our ears before we open our eyes. We hear the world before we see it. The event of the truth of the world is the event of the resounding of the world. We do not dwell in the silent house of being, but in the tower of sound that is the worlding of words, the cacophony of voices, things, animals, machines and bodies sounding the world. The world is the tympanum of the sounding of our somatological technologies.

REFERENCES

Appelbaum, David (1990). *Voice*. Albany, NY: SUNY Press.

Carson, Anne (1995). *Glass, Irony, and God*. New York: New Directions Books.

Cavarero, Adriana (2005). *For more than one Voice: Toward a Philosophy of Vocal Expression*, trans. Paul A. Kottman.Stanford, CA: Stanford University Press.

Dolar, Mladen (2006). *A Voice and Nothing More*. Cambridge, MA: The MIT Press.

Heidegger, Martin (2012). *Contributions to Philosophy (of the Event)*, trans. Richard Rojcewicz and Daniela Vallege-Neu. Bloomington, IN: Indiana University Press.

_____. (2010). *Being and Time*, trans. Joan Stambaugh, revised with a Foreword by Dennis J. Schmidt. Albany, NY: SUNY Press.

Ihde, Don. (2010). *Heidegger's Technologies: Postphenomenological Perspectives*. New York: Fordham University Press.

_____. (2009). *Listening and Voice: A Phenomenology of Sound*, 2nd edition, revised and expanded. Albany, NY: SUNY Press.

_____. (2001). *Bodies in Technology*. Minneapolis: University of Minnesota Press.

Garret Keizer, Garret (2010). *The Unwanted Sound of Everything We Want*. New York: Public Affairs.

LaBelle, Brandon (2010). *Acoustic Territories: Sound Culture and Everyday Life*. New York and London: Continuum.

Mendieta, Eduardo (2014). The Sound of Race: The Prosody of Affect in *Radical Philosophy Review* Vol. 17, No. 1 (2014): 109–131.

_____. (2012). What can and cannot be rescued—taking leave of Heidegger's Hut: On Don Ihde's *Heidegger's Technologies: Postphenomenological Perspectives* in *Philosophy and Social Criticism*, Vol. 38, No. 2 (2012): 227–233.

_____. (2012a). From Metaphysical Racism to Biopolitical Somatology in *Hypatia* 27. 4 (2012): 906–911.

_____. Bodies of Technology: On Don Ihde's *Bodies in Technology* in *Journal of Applied Philosophy: Journal of the Society for Applied Philosophy*, Vol. 20, No. 1 (2003): 95–101.

Plessner, Helmut (1981). *Die Stufen des Organischen und der Mensch. Einleitung in die philosophische Anthropologie. Gesammelte Schriften, IV*. Frankfurt am Main: Suhrkamp Verlag.

Index

Aalborg, 92, 95, 99
Abney, Keith, 174–76
Actor Network Theory, 11, 164, 171, 192, 213
Adams, Vincanne, 65
Afghanistan, 214
agency, moral agency, 3–4, 6, 9–11, 169, 179, 182, 184, 186, 190–91, 196, 206
Alac, Morana, 181–82
Alba, 155, 158n1
Alcoff, Linda, 94
alterity, 9, 96, 132–33, 171–72, 177, 185
amodernism, amodern, 10–11, 202
analytic philosophy, 2–5, 7–9, 11–12, 241
Anderson, Warwick, 65
Anette, Forss, 130, 142–43
Apel, Karl-Otto, 13n6, 242
Aristotle, 150
artificial intelligence, 58, 169
Asimov, Isaac, 174–76
Athena, 153
Australians, 62
autonomy, 5, 10, 165, 199
Auto-Tune, 118

Bachelard, Gaston, 96
Bacon, Francis; Baconian, 5, 212
Bain, Erica, 207
Baird, David, 194
Barad, Karen, 191
Bassala, George, 48, 50, 52n14
Baudrillard, Jean, 151, 153, 158n3
beatport, 116, 118
Beck, Urlich, 215
Beetles, 118
Beijing, 235
being-in-the-world, 40–41, 43, 51
Bell, Alexander Graham, 42
Bennett, Jane, 179
Bernard, Claude, 94
Besmer, Kirk, 131, 134, 137
Bhopal, 58
Bhopal, India, 215
Bijker, W., 4
Bioart, 150, 155–56
Birkler, J., 163
body schema (skhema), 86–87, 99, 136
Borch, Martin, 91
Borgmann, Albert, xiv, 4, 129, 139, 158n3
Branagh, Kenneth, 83
Breazeal, Cynthia, 174
Bristol, England, 120, 123
Brothers Grimm, 39
Bruun, Maja Hojer, 186n2
Burke, James, 58

Cage, John, 122
Calder, Peter, 232, 236
California, 108, 110–11, 120
Callon, Michel, 4
camera, vii, 26, 31, 47, 50, 83, 111, 122, 142, 162, 232, 235
camera obscura, vii
Carshalton, Surrey, 166
Cartesian, xiii, 2, 12, 22, 93, 99, 106, 185, 226–28
Carusi, Annamaria, 34, 36n3
cell phone, xiv, 39–48, 50, 52, 139, 247
Ceruzzi, Paul, 41
Chakrabarty, Dipesh, 64–65
Chancellor A. G. Moltke, 90
Chernobyl, 211, 215, 217
China, 214
Collabinator, 125
Colombia, 246
colonialist, 60
Colorado, 205
compass, xi–xii, 62–63
computers, 25, 27, 41, 46, 74–75, 79, 88, 95, 117, 123, 130, 133, 140, 246
constructivist, 212–13
continental philosophy, continentalism, continentalist, 4, 7, 240–42
Conversation Analysis, 192, 195
Copenhagen, 90, 95, 110
Crimea, 161
Cubase, 122

Dasein, 243
Davidson, Donald, 7
Deckadance, 119
De Gama, Vasco, 61
De Mul, Jos, 200
Denmark, 90, 110, 170–71, 180, 186n1
De Preester, Helena, 130, 134–36
Descartes, René, vii, 5, 13n5, 107, 226–27
determinism (social, technological), 57–59, 66, 189, 212
De Thurah, Lauritz, 90
Devi, Mahasweta, 60
Dewey, John, vii–viii, 87, 149
Dilthey, Wilhelm, 7–8, 13n7, 14n18
DJcity, 124, 126
DJ Merc, 121, 124–25, 126n1
DJ TIMarbury, 123
Dobbs, David, 205, 209
Dodge and Fuski, 120–23
Drones, 26, 28, 30–33
Drummond, John, 167
dualism, 5, 106, 227
Dubai, 233
dubstep, 115–18, 119–23, 126
Dub Turbo, 118
duck-rabbit, xiii–xiv

Ealing, 29
Eaton, Richard, 65
echo-location, vii–x
Eco, Umberto, 152–53
Egypt, 30, 51n1
Eigtved, Nicolai, 90
Electric Dance Music (EDM), 115, 119, 120, 122
Ellul, Jacques, 59–60
Elvis, 118
embodiment, viii–xii, 6, 9, 23, 31, 44–45, 48, 57, 59, 62–64, 66, 71–74, 86, 88–89, 95, 112–13, 122, 131–37, 150–55, 157, 172, 183, 208, 241, 244–47
empirical turn, xiv, 1–2, 10–12, 14n15, 57, 193
empiricism, 2, 6–8, 12
enframing, 12, 96, 98
Engels, Friedrich, 40–41
enlightenment, 10, 60, 66
epistemology, vii–viii, 3, 5, 7–8, 11–12, 13n8, 14nn13, 15, 15n23, 131, 145, 194, 221, 239
essentialist, 149, 208
ethics, 3–4, 6, 8–9, 11–13, 15n23, 28, 87, 169–76, 178–80, 182–86, 189–90, 193, 196, 198, 212, 217, 223, 239
Europe, European, xi, xiv, 4, 7, 52n5, 57, 61–63, 65, 91, 174, 208, 214, 226
externalist, 4, 6

face-vase, xiii–xiv
Federer, Roger, 233
Feenberg, Andrew, 4, 8–9, 129, 158n1, 189
feminist, 60, 64, 136, 145
Ferraris, Maurizio, 46
Feyerabend, Paul, 60
Figal, Günter, 153
Fischer, John M., 22
Fischetti, Mark, 236n2
Florence Nightingale, 161
FL studio, 119
Foster, Jodie, 207–8
Foucault, Michel; Foucaultian, 8–10, 40, 42, 48–50, 64, 85, 93–94, 245
Frederiksstad, 90
FruityLoops, 121
Fukushima, 211–12, 214–15, 217–19
Fuller, Steve, 13n8

Gadamer, Hans-Georg, 13, 87
Galen, 226
Galileo, xii, 7, 59, 107, 161–62, 216
Galison, Peter, 194, 241
Garcia, Andy, 83
Gauthier, M. P., 91
Geiger counter, xi, 230
Germany, German, 52n9, 233, 246
Gibson, James-Jerome; Gibsonian, 225, 227
Gilbert and Sullivan, 175
Glaser, Donald, 191
Goggin, Gerard, 47
Gomes, Alaric, 233
Google, xi, 25, 197–98
Guha, Ranajit, 64

Hacking, Ian, 191
Hanghøj, Signe, 186n2
Hantula, Richard, 52n10
Haraway, Donna; Harawayan, 24, 34, 62, 64, 129, 145, 156, 241
Harding, Sandra, 241
Hardy, A., 92
Harman, Graham, 51n1, 202
Harvey, William, 226
Hasse, Cathrine, 130–31, 138, 141, 143
Hayles, Kathrine, 179
hearing aid, x, 27, 244
Hegelian, 50
Heidegger, Martin; Heideggerian, Heideggerese, 1, 5, 7, 9, 11, 13n7, 14n11, 40–42, 50, 51n2, 58–59, 105–7, 129, 132, 139, 142, 149–50, 153–55, 157, 202, 212, 221, 239, 240–44
Herfel, Roland, 233
hermeneutic, hermeneutics, hermeneutical, xii–xiii, 2, 7–9, 12, 13n6, 14n11, 20, 23–24, 29, 35, 45, 48, 57, 64, 66, 84, 106–7, 113, 130, 132, 133, 135, 137, 141, 143–44, 151–52, 172, 177, 191, 194–95, 202
Hipstamatic, 29
Hiroshima, 58
Hoel, Aud Sissel, 34, 36n3
horizon, xi, 21–22, 42, 51, 71, 154, 241, 245
Horkheimer, Max, 212
humanist, 4, 6, 10–11, 163, 173, 192, 207
Hume, David; Humean, 6
Husserl, Edmund; Husserlian, vii–viii, xiii, 13n7, 15n19, 19–24, 27, 35–36, 59, 105–7, 138–39, 150, 161, 202, 230, 233, 241–42

Ihde, Don, xiv, xvii–xviii, 1, 4, 6–10, 12, 13nn6–8, 14–15n19, 19–20, 23–24, 29, 35, 40–43, 57–59, 62–66, 69–73, 75–78, 80n3, 87–89, 101, 106, 107, 112–13, 115–17, 119–20, 129–34, 136–43, 145, 149–57, 161–65, 169–72, 177, 183, 190, 192, 194, 206, 207, 209, 239–45, 247
India, Indian, 60–61, 65, 215
Instagram, 28–30
instrumentalism, instrumentalist, 149, 205–6
instrumental realism, 106, 194
intentionality, x, 10, 116, 134, 202, 228
internalist, 4

iPod, 246–47
Italy, 92

Japan, Japanese, 174, 179, 211–12, 214, 217–19, 222–23
Jaspers, Karl, 200
Jonas, Hans, 50, 98
Julaha, 65
Jutland, 92, 98–99, 186nn1–2
Jutland, Denmark, 92, 98–99, 186n1

Kac, Eduardo, 150, 155–56
Kansas, xiii
Kant, Immanuel; Kantian, 113, 193, 239
King Frederik V, 90
Koopman, Colin, 48–50, 52n13
Kripke, Kripkean, 3
Kroes, Peter, 189
Kuhn, Thomas; Kuhnian, 14n18, 231
Kurokawa, Kiyoshi, 218–19, 223
Kurzweil, Ray, 173

La Mettrie, Julien Offray de, 226, 230
Latour, Bruno; Latourean, 4, 8–9, 60–61, 162–63, 172, 177, 191–92, 202, 207–9, 241
Laudan, Larry, 14n13
Le Corbusier, 98, 100
Leder, Drew, 85
Levinas, Emmanuel, 171
Levinson, Paul, 50
lifeworld, 35, 57, 59, 63, 65–66, 69, 71, 73–74, 84–85, 106–7, 112, 171, 230–31, 233–34
Live and Push, 119
lived body, 59, 85, 244
Locke, John, vii
logical Empiricism, 2, 7–8
Logic Studio, 118
London, 29, 117, 235

magnifying glass, 76
Manhattan, xvii
Marconi, Guglielmo, 42
Marcuse, Herbert, 59–60, 101
Mars, 142
Marx, Karl, Marxist, neo-Marxist, 4, 40–41, 65–66
material hermeneutics, 64, 84, 107, 172, 177, 194
McEnroe, John, 232
mediation, xiii, xv, 6, 8–9, 11–13, 13n6, 14n19, 100, 106n5, 107, 110–11, 117, 120, 133, 166, 172, 177, 190–202, 208
mediation theory, xv, 192, 198–202
Mercury, 175
Merleau-Ponty, Maurice, viii, x–xi, xiii, 36n3, 59, 64, 86–87, 99, 102, 105–7, 132, 202, 226–30, 245
metaphysics, 149–50, 192–94, 199, 201, 241–42, 247
Michaud, Yves, 230
microscope, x, 106–7, 162, 164, 191
Miki, Kiyoshi, 219–20
Mishra, Pankaj, 61
Mitcham, Carl, 3–4, 13n2
modernism, 10–11, 85, 100, 202
Moore, Gordon, 41–42
moral agency, 3–4, 6, 9–11, 169, 179, 182, 184, 186, 190–91, 196, 206
multistability, multistable, xiii–xv, xvii, 1, 19–20, 23–24, 29, 69–70, 72–80, 80n2, 84, 89, 101–2, 106n6, 112–13, 117, 120, 130–31, 134, 137–45, 149–50, 155, 169, 172–73, 176–80, 183–85, 206

Nadal, Rafael, 233, 236
Nagasaki, 58
Nagel, Thomas, vii–ix, xiv
Nanda, Meera, 58
Nandy, Ashis, 60
Napster, 123
NASA, 173, 216, 219
National Rifle Association (NRA), 205–7
Navrátilová, Martina, 229
Necker cube, xiii–xiv, 138, 143
Nevada, 31
Newgrounds, 123, 125
New Guineans, 62
New York, 121, 246

Nietzsche, Friedrich, 19
night vision goggles, x
Nishida, Kitaro, 212, 218–22
nonfoundationalism, 139
North America, North American, 1–2, 8, 12, 65

objectivity, vii, 117, 201
O'Callaghan, Jonny, 233
Odense, 95, 110
ontology, 10, 13n8, 34, 36, 94, 100, 221, 228, 240–41

Palo Alto, 28
Pandey, Gyan, 65
Paris, 91
Paro (robot), 170–71, 173, 178–79, 181, 183–85
Parthenon, 153, 155
Pennsylvania, 121
Perrow, Charles, 215–16
Pfeifer, Jessica, 14n18
philosophy of science, 2, 7, 14n18, 107
philosophy of technology, xi, xv, 1, 3–5, 12, 13n2, 42, 51, 57, 69, 75, 129, 150, 167, 189–90, 192–93, 196, 201, 212–13, 219–22, 240
physicalist reductionism, vii–viii
Piccinini, Patricia, 150, 156–57
Pickering, Andrew, 191
PictureShow, 29
Pistorius, Oscar, 235
Plant, Sadie, 46
Plessner, Helmut, 244
polymorphism, xiv
positivism, positivist, 2, 4, 7–8, 11, 13, 14nn11–13, 15n23
postcolonial, postcolonialism, xv, 57, 59–61, 64–66
posthumanism, posthumanist, posthuman, 10–11, 169–70, 173, 176, 178–80, 182, 184–85
Postman, Neil, 41
postmodernism, postmodern, postmodernity, 50, 101, 111, 149–51, 153, 157, 179, 201
postphenomenology, viii, x, xi–xiv, xvii–xviii, 1–2, 6, 11–13, 14n19, 19, 26, 42–44, 49–51, 62–63, 66, 69–70, 72, 79, 84, 105, 106n10, 107–8, 112–13, 129–31, 134, 139–41, 144–45, 149–50, 157, 192, 194–95, 201
postpositivism, postpositivist, positivistic, 2, 4, 8, 12, 13n1, 14nn13, 19, 28
pragmatism, pragmatist, pragmatic, vii, 2, 12, 57, 61, 64, 83, 87, 106, 108, 129, 137, 139
Prakash, Gyan, 61
pregiven, 34, 194
pre-reflective, prepredicative, 150, 157, 231
present-at-hand, presence-at-hand, 46, 48, 51n2
presentism, 177, 179, 185
Primo, primo-posthuman, 173, 180, 184
Pritchard, Sara, 215
projection, 150, 153, 157
propellerhead, 119
Pulawateans, 63, 66

Quine, W.V.O.; Quinean, 2, 7, 14n11

Rabinow, Paul, 50
radio, xi, 39, 41–42, 52n9, 95, 116, 124, 134, 208
Rao, Narayan Velcheru, 61
Ravizza, Mark, 22
ready-to-hand, 45, 51n2
renaissance, vii, 101
representational, 25
Ricoeur, Paul, 241
Rolling Stones, 118
Rorty, Richard, 241
Rosenberger, Robert, 75–78, 169, 183
Rouse, Joseph, 7–9, 13n9, 14nn11, 18

Salt Lake City, Utah, 111
San Diego, 181
Sarkar, Sahotra, 14n18
Sausserian, 151

Scharff, Robert, 64
sedimentation, sediments, sedimented, 23, 50–51, 136, 143, 230, 239
Segerink, Loes, 197
Selinger, Evan, 35n1, 66, 72, 78
Shimomura, Osama, 158n5
Shulman, David, 61
Sierra Nevada, 109
Silbot (robot), 171, 173, 177–85
singularist, 169, 173–74, 176–79, 183–85
Sjælland, 110
Skrillex, 118
Skype, 247
SoundCloud, 118, 121, 125–26
South Africa, 235
South Korea, 171, 180, 184
Speedy (robot), 175–76, 184
Spinoza, Baruch, 178
Spivak, Gayatri, 64
Stern, Nick, 28
Stony Brook, xvii, 1
subjectivity, vii–viii, 20, 26, 105, 117, 201, 228
subnav, 118
Subrahmanyam, Sanjay, 61
substantivism, substantivist, 149
surveillance, 33
Sweden, 163–64

Tahir Square, 30
Takeno, Junichi, 174
Tansey, E. M., 92
Taylor, Charles; Taylorism, 7–8, 13n9, 14n12, 94
technoscience, technoscientific, xvii, 1, 4, 6–7, 9–12, 57–58, 62, 64–65, 106, 172, 240
telephone, 39, 42, 52n9, 247
telescope, x, xi–xii, 7, 43, 105–8, 112–13, 142, 152, 154–55, 161–63
television, 27, 39, 41, 95, 208, 247
Tenner, Edward, 215
Terada, Torahiko, 214, 219, 222
Teyssot, Georges, 94
thermometer, xii, 164
Three Mile Island, 215, 217
thrownness, 157
Tibet, 214
Tohoku, 211
Tosaka, Jun, 219–20
transhuman, transhumanist, 173
truth, 13n9, 21–22, 26, 85, 107, 116, 149–50, 153, 155, 157, 232, 241, 243, 247
Tsuruga, 217
Turkle, Sherry, 179
turntables, 117

U2, 118
United Kingdom, UK, ix, 118
United Nations, 44
United States, US, 14n18

van den Berg, Albert, 197
Vanilla Ice, 125
variational theory, variational analysis, xiv, 57, 72–73, 75, 120, 138–40
Vattimo, Gianni, 12
Verbeek, Peter-Paul, 4, 6–7, 9–12, 34, 76–79, 117, 131, 134, 139, 169, 172, 174
Vermaas, Pieter, 189
Vermont, ix
Veruggio, G., 174
Vinge, Vernor, 173
Vita-More, Natasha, 173
Voege, Debra, 52n10

Walkman, 246
Warhol, Andy, 151
Wellner, Galit, 139
White, Lynn, 58, 214
Whyte, Kyle Powys, 131, 138, 140
Wimbledon, 233

X-Ray, xii, 91, 162

YouTube, 118, 125, 174
Youzhny, Mikhail, 233

Zabala, Santiago, 12
Zen Buddhism, 222

About the Contributors

Robert P. Crease is a professor in the Department of Philosophy at Stony Brook University, New York. He has written, translated, or edited over a dozen books on history and philosophy of science. He is the author of "Critical Point," a monthly column for Physics in Perspective, and Co-Editor-in-Chief of Physics in Perspective.

Jan Kyrre Berg Olsen Friis is Associate Professor of Philosophy of Science and Technology at Copenhagen University. He has authored, edited and co-edited numerous books on philosophy of technology, philosophy of time, metaphysics, and science. He has published papers on topics such as hermeneutics; perception; philosophy of technology; time concepts in physics, experience, and measurement.

Don Ihde is Distinguished Professor of Philosophy, Emeritus, Stony Brook University, New York, US. He is the author of twenty-one books as single author and editor or co-editor of many more. Recent books include *Postphenomenology and Technoscience* (2009), expanded second editions of *Listening and Voice* (2007), and *Experimental Phenomenology* (2012). Ihde has lectured internationally and works of his appear in translation in fifteen languages.

Robert C. Scharff is Professor of Philosophy Emeritus at the University of New Hampshire. He is author of *Comte After Positivism* (Cambridge, 2002) and *How History Matters to Philosophy: Reconsidering Philosophy's Past After Positivism* (Routledge, 2014); co-editor (with Val Dusek) of *The Philosophy of Technology: The Technological Condition* (2nd ed.,

Blackwell, 2014); and former editor of *Continental Philosophy Review* (1994–2005). He is currently finishing a manuscript, *Inheriting Technoscience*.

Srikanth Mallavarapu has a PhD in English from SUNY Stony Brook, where he also was a member of Don Ihde's Technoscience Research Group. His research explores the negotiation of modernity in the postcolonial context. He has been a Brittain Postdoctoral Fellow at Georgia Tech and a SAGES Fellow at Case Western Reserve University. He currently teaches in the English department at Roanoke College, Virginia.

Shannon Vallor is Associate Professor of Philosophy at Santa Clara University in Silicon Valley, where she explores the moral, epistemic and phenomenological implications of emerging science and technology, and in particular, the impact of technology on the cultivation of human virtues. She is Vice-President and President-Elect of The Society for Philosophy and Technology (SPT).

Galit Wellner is a post-doc researcher at Ben Gurion University of the Negev, Israel. Galit studies digital media and its interrelations with humans. She graduated the STS program at Bar-Ilan University. In her dissertation she explored the cell phone as a case study for twenty-first-century technologies using Postphenomenology. Before her PhD, she worked for many years for the Israeli high-tech industry in various positions.

Kyle Powys Whyte holds the Timnick Chair in the Humanities in the Department of Philosophy at Michigan State University and is a faculty member of the Environmental Philosophy & Ethics graduate concentration. His primary research addresses moral and political issues concerning climate policy and Indigenous peoples and the ethics of cooperative relationships between Indigenous peoples and climate science organizations. He is an enrolled member of the Citizen Potawatomi Nation.

Lars Botin is Associate Professor at Department of Development and Planning, Aalborg University (DK). He has during the past decade investigated the intersection in between art, science and technology in the domain of health care, namely health informatics and hospital design. He is the instigator of a new research-domain: techno-anthropology in postphenomenological perspective.

Michael Funk works as Research Assistant at the Institute for Philosophy of Technology at TU Dresden (Germany). His current research includes intercultural and transdisciplinary philosophy of technologies and sciences, with special interests in robotics, life sciences, paleoanthropology, genetic engineering, synthetic biology and music. www.funkmichael.com

Stacey O. Irwin is Associate Professor at Millersville University of Pennsylvania, where she teaches a variety of digital media production, writing, and theory classes. Her interest in studying media through philosophy began when she started teaching students to use sophisticated media production editing and design tools in the classroom. This blossomed into critically studying the experiences of "making and doing" media. Her work has been published in Techné: *Research in Philosophy and Technology*, *Tamara Journal for Critical Organization Inquiry*, *Explorations in Media Ecology*, and *Human Studies*, and she is currently working on her first book. She is also a documentary producer and has spent time working in radio and television.

Robert Rosenberger is an Assistant Professor of Philosophy at the Georgia Institute of Technology. He works primarily in the development of postphenomenological theory, with case studies addressing topics such as Mars satellite imaging, cell-phone-related driver distraction, sample freezing techniques in neurobiology, educational frog dissection simulations, and public space designs that discriminate against the homeless. He is Editor-in-Chief (and co-founder) of the book series "Postphenomenology and the Philosophy of Technology," of which this volume is a part, and co-curator of the website postphenomenology.org

Roisin Lally is teaching in Gonzaga University. She is a PhD candidate at National University Ireland Galway. Her research interests include philosophy of technology, metaphysics, contemporary philosophy, and the history of philosophy.

Anette Forss is a Senior Lecturer in Nursing at the Karolinska Institutet, Sweden and a Visiting Scholar at the Department of Philosophy, Stony Brook University, SUNY (New York), US. Her interest is to merge philosophy of technology (postphenomenology) and nursing. She is the principal investigator of an ethnographic study that explores the nurse-technology-patient interface in a variety of oncology related health care regimes. She has co-authored papers published in a variety of scientific journals, authored a chapter in *Medical Technologies and the Lifeworld*, Routledge (2007), and is co-editor, with Chris Ceci and John Drummond, of *Philosophy of Nursing—5 Questions,* Automatic Press/VIP (2013).

Cathrine Hasse is Full Professor at Aarhus University in Denmark at the department of Education. She is leading the research group Future

Technologies, Culture and Learning engaged in research on educational technologies and global cultural learning processes. She was educated as an anthropologist, and has also studied physics education and physics research. She did her PhD on "cultural learning processes" in a physics institution where she followed a group of young male and female physicist's students in their first year of study. This project developed into a longitudinal study where she has followed the same group of students more than 6 years. In her next projects "The cultural Dimensions of Science" and UPGEM, an EU project, financed by EU 6th framework program, she and her research groups studied cultural norms in the sciences. In her work she takes a special interest in the relations between technology, culture and learning. She is the author of, for example, *An Anthropology of Learning*, published by Springer, an MC in the COST action New Feminist Materialism and active participant in the 4S network of Science and Technology Studies with several workshops on postphenomenology and anthropology.

Peter-Paul Verbeek is Professor of Philosophy of Technology at the University of Twente and president of the Society for Philosophy and Technology. His research focuses on human-technology relations in relation to knowledge, morality and metaphysics. Among his publications are *Moralizing Technology: Understanding and Designing the Morality of Things* and *What Things Do: Philosophical Reflections on Technology, Agency, and Design.*

Evan Selinger is an Associate Professor at Rochester Institute of Technology, where he also serves as the Head of Research, Communications, Community & Ethics at the Center for Media, Arts, Games Interaction & Creativity (MAGIC). He has written extensively about core issues in the Philosophy of Technology.

Junichi Murata studied at the Department of History and Philosophy of Science, the University of Tokyo, is presently Professor at the Department of Philosophy, Rissho University and Professor Emeritus of the Tokyo University. His major book publications include: (in English) *Perception, Technology, and Life-Worlds* (Collection UTCP 1, 2007).

Shoji Nagataki is Professor of Philosophy at the School of International Liberal Studies, Chukyo University. His research interests includes phenomenology, philosophy of embodiment, philosophy of science, and cognitive robotics. He is author of "Phenomenlogy and the third generation of cognitive science" (in *Human studies*, 2007) and co-author of *The Evolution of Social Communication in Primates* (Springer, 2014).

Eduardo Mendieta is professor of philosophy at Penn State University. He is the author of *The Adventures of Transcendental Philosophy* (Rowman & Littlefield, 2002) and *Global Fragments: Globalizations, Latinamericanisms, and Critical Theory* (SUNY Press, 2007). He is also co-editor with Jonathan VanAntwerpen of *The Power of Religion in the Public Sphere* (Columbia University Press, 2011), and with Craig Calhoun and Jonathan VanAntwerpen of *Habermas and Religion* (Polity, 2013), and with Stuart Elden of *Reading Kant's Geography* (SUNY Press, 2011) He recently finished a book titled *The Philosophical Animal*, which will be published by SUNY Press in 2016.